Contents

Contents

Developments

in

Mathematics Teaching

F. R. Watson

Open Books
London

First published in 1976 by Open Books Publishing Ltd,
87–89 Shaftesbury Avenue, LONDON W1V 7AD

© F. R. Watson 1976

Hardback: ISBN 0 7291 0085 5   *0047589*

Paperback: ISBN 0 7291 0080 4

90749

Text set in 11/12 pt Photon Imprint, printed by photolithography, and bound
in Great Britain at The Pitman Press, Bath

374.51
WAT

Editor's introduction

The theme of this series of books is the changing classroom. Everyone knows that schools change – that despite all the influences of tradition things aren't the same as they used to be. Yet during the past decade the change has been on such an unprecedented scale that in many ways schools have become surprising places not only to those who work with them – like parents and employers – but even to those who work in them, like teachers and students.

There are many reasons for these changes. Some are organisational, like the move to comprehensive secondary schooling, the raising of the school leaving age, new pre-school classes, and 'de-streaming', where children of all abilities work together. But even more spring from the way the teacher works in the classroom – from the increasing emphasis on individual methods, on creativity rather than remembering, on new patterns of assessment and examination, and on the use of a wide variety of project methods.

Such changes have certainly transformed the life of many classrooms and made school a different place for teachers and their students. This series is about life in those classrooms, for it is here that we can not only see change but understand it and get to grips with its effects on young people and on the society in which they will live.

In this volume F. R. Watson writes about new developments in the teaching of mathematics. The changes in mathematics have probably been the most visible and the most discussed of all the changes in the curriculum. Many parents, employers, and even teachers of other subjects, are bewildered by new and previously unfamiliar topics such as sets, symmetry and multi-base calculations. From the use of the new apparatus in the infant school to the new O level examination papers, the evidence of radical change is unmistakable. F. R. Watson, a mathematician with many years' teaching experience, presents a coherent map of the new mathematics, explains its problems and analyses its objectives with a wealth of examples of the work of the schools, and of the new projects and programmes through which the subject continues to develop.

John Eggleston

Acknowledgements

The author and publishers would like to thank the following for permission to reproduce material used in this work:

W. and R. Chambers, John Murray and the Nuffield Foundation for the extracts from Nuffield Mathematics Project, *Computation and Structure 3* (1968), from *I Do and I Understand* (1967), and from *Guide to the Guides* (1973); Chatto and Windus Educational (Granada Publishing) for the extract from *Mathematics for Outdoors* (Schools Council Publications); M. Ruth Eagle for the basis of the diagram in chapter 7; Heinemann Educational Books for material from Mathematics Applicable, *Logarithmic/Exponential* (1975); the Joint Matriculation Board for the question on p. 43; B. J. Lang for material on p. 118; Oxford University Press for material from K. L. Gardner, J. A. Glenn and A. I. G. Renton, *Children Using Mathematics* (1973); Schofield and Sims for material on p. 90; Schofield and Sims and Schools Council Publications for material from Mathematics for the Majority Continuation Project, *Buildings* (1974); Schools Council Publications for material from *Newsmaths* 6/7 (1973), and from Sixth Form Project *Newsletter* 4 (1971).

I am grateful to many people who have helped in various ways in the production of this book. A special word of thanks should go to Mrs M. R. Leedham and Mrs J. E. Simpson, whose skill and speed in deciphering the first draft were truly remarkable.

F.R.W.

PART I

1 Introduction: the catalysts of change

'Modern' mathematics

Much has already been written about mathematics teaching in the last fifteen years, a period which has seen considerable changes right through the age range from age five (or even three?) through to tertiary education. In primary schools the widespread abandonment of eleven plus selection has removed one constraint and permitted teachers in many areas to take a wider view of mathematics. As a consequence, computational slickness in arithmetic lost its former importance and instead work was introduced on geometry (as 'shape' – i.e. area, similarity, and symmetry) and on algebra (graphs as 'pictorial representation', equations, and set-notation). In the secondary field, projects proliferated – regions decided to 'go it alone', enterprising heads of department produced their home-grown syllabuses, and no publisher, it seemed, could afford to be without a 'modern maths' series. The Mathematical Association (1968) listed some twelve mathematics projects in British secondary schools (St Dunstan's, Manchester, Mathematics in Education and Industry, Midlands Mathematical Experiment, Nuffield 5–13, Psychology and Mathematics, Royal Liberty School, School Mathematics Project, Mathematics for the Majority Project, Scottish Mathematics Group, Shropshire, Swansea). Since then several others have been added. Among these are: the Mathematics for the Majority Continuation Project (chapter 11), Sixth Form (chapter 12), 11–16: a critical review, Welsh Medium School, Primary Evaluation (p. 69) and Early Experiences – all supported by the Schools Council – together with Continuing Mathematics (p. 126), S.M.P. Middle School (p. 70), Fife (p. 26), SMILE (see Gibbons 1975), Hertfordshire (p. 26), Kent (see Banks 1975), and other localised activities. This is not an exhaustive list. See also figure 7.1.

With such a wide diversity of ages, interests and concerns involved, it is hardly surprising that the term 'modern mathematics' quickly became a portmanteau expression for a whole host of developments. ('When I use a word, it means just what I choose it to mean,' as Humpty Dumpty said.) It thus became a rallying cry for those who wished to attack (or defend) the *status quo*, a boon to publishers (and to headmasters on speech day), and a whipping boy for industrialists complaining of the quality of their recruits, as well as a label for the very great deal of planning, re-thinking, experimentation, writing, and sheer hard work which has gone into the series of major developments.

Yet this surge of activity has not brought the educational utopia which some (rather naively) expected; with hindsight one can see that some of the early hopes were unduly optimistic. '. . . the whole attitude to the subject can be changed and "Ugh, no, I didn't like maths" will be heard no more' (Nuffield Foundation 1967). Indeed in some ways the changes have served only to highlight the problems and clarify the questions which must be answered before curriculum reform in mathematics can begin to produce the desired outcomes. We have underestimated the magnitude of the task and the difficulty of bringing about change; the educational system as a whole has high inertial mass! It is not simply a matter of changing the textbooks – many other questions must be considered:

What is the purpose of teaching mathematics in school?

How much – and what sort of – mathematics does the average 'man on the Clapham omnibus' need to know?

What types of mathematics are needed to service such subjects as physics, chemistry, engineering, biology?

Will the growth of mathematically oriented studies in new subjects

such as economics, geography, geology, psychology, sociology
make different demands? Is there a core syllabus?

Who will teach all this material?

Are there major ideas in mathematics? What are they and are they
changing?

Is there a place for mathematics as a general cultural subject?

What is the nature of mathematics?

How is mathematics done? How is it learned?

Are these the same — or similar?

How do we teach pupils to be 'self-reliant' in their mathematical
thinking?

Shall we proceed from applications to mathematics or vice versa?

How do we teach the skill of applying mathematics?

How do we motivate pupils — how do we make mathematics 'relevant' for them?

How do we train mathematics teachers — or re-train them — for their
changing tasks?

This is a formidable list of questions, to which must be added others
concerned with more general teaching problems:

How do we cope with mixed ability classes, or with individualised
learning by pupils?

and even

What is education for?

This book falls into three parts. First comes a brief description of
the current position, which tries to identify the areas of concern to
which curriculum developments have been a response. This is
followed by a discussion of some case studies, on primary, middle
secondary, and sixth-form mathematics projects, together with an
example of a less institutionalised activity. These illustrate in concrete and specific terms the issues already raised. The final part discusses some of these issues further and attempts to identify some
directions for future progress. Inevitably, this section is subjective
and personal, for crystal-ball gazing is a dangerous occupation unless

one is skilled at avoiding a committed prediction, thus giving no hostages to fortune!

The teacher

Any discussion of curriculum development in mathematics sooner or later comes up against the problem of the *status quo* and the limitations which circumstances impose on the actions of individual teachers. Their task is often so circumscribed by the situation in which they find themselves that their control over curriculum, examinations, facilities, and even teaching method appears minimal. What use is airy speculation about the purposes of teaching mathematics, the nature of mathematical activity, the aesthetic pleasure derived from its study, the desirability of involving pupils in initiating as well as solving problems . . .?

Though dramatic 'liberations' are unlikely, gradual developments do take place — a little here, a little there. Unless they know which way they would like to go, how can teachers influence the direction of change? Further, teachers may have more control over the situation than they realise; some obstacles are not so immovable as they first appear. As an example, if the examination syllabus which 'they' have imposed is felt to be a constraint, a Mode II examination (externally administered on the school's own syllabus) may be used, or if independent project work is to be encouraged, a Mode III examination (conducted by the school and externally moderated) is an alternative. Teachers undoubtedly feel strongly about their responsibility to their pupils and wish to do their best by them; the children, their parents, and their future employers all have a stake in examination success. But teachers are not obliged to distort their teaching to 'coach for the examination'. To resign oneself to the system as it is may appear 'realistic', yet the result is a 'dead' approach, wearisome to teacher and pupil alike. It has often been remarked that an important by-product of much curriculum development, especially where it has involved groups of teachers meeting for discussion, has been the freshness and enthusiasm generated by 're-thinking' some of the basic questions of mathematics teaching. Change is not beneficial of itself, and the organisational upheavals in secondary education (and more recently in the tertiary field) have inevitably diverted a tremendous amount of energy from the fundamental tasks of teaching. One hopes that pulling up institutions by the roots every few years will not become a continuing feature of the way our educational system is administered.

Yet it is clear that there are possibilities for initiative in curriculum development by teachers at the local level – and if these are not taken up there is a serious danger that curriculum planning may become more centralised 'in the interests of efficiency', and the system, by stifling the non-conforming, may reduce itself to monolithic mediocrity. During a recent visit to the U.S.S.R. to study mathematical education, I asked a question about centralised curriculum planning. The substance of the reply was: 'We think about the problems very carefully, decide what to do, and all get it right together'.

In many countries curriculum planning is undertaken centrally. Consideration of developments in mathematics teaching both here and in other countries during the past two decades leads inevitably to the conclusion that the problems are too large and complex for any committee, however eminent and however able its members, to get the answers right first time. However, if teachers are to be active participants in curriculum development, rather than the passive recipients of courses of action prescribed by authority, they must be willing to probe, to question, and examine their own teaching, their objectives and strategies, prepared to keep abreast of new developments and learn from the experience of others, constantly rethinking the teaching of mathematics in their own circumstances.

Catalysts of change

Looking back over the last fifteen years it is of interest to ask what factors brought about the changes in the mathematics curriculum. It is said that school mathematics in 1955 was little different from that of the 1930s; the subject had assumed a settled, almost fossilised, appearance. As was the case in science, new developments at the 'growing edges' of the subject were occurring rapidly, and whole new areas of mathematics were being explored. Particularly great changes were occurring in university mathematics. Ought these not to be reflected in some way in the syllabuses used in schools? In reality this is not as strong an argument as might be supposed; the 'modern' topics in today's mathematics syllabuses mostly have their origin in the pre-1900 era. Nevertheless the argument of the obsolescent syllabus was strongly advanced and it was suggested that pupils' work in schools should include material reflecting the changes which had taken place in the advanced study of mathematics.

There were other educational reasons for reform, such as the growth of knowledge in developmental psychology and its implications for the teaching of mathematics – but whatever the reasons, as is often the case it was economic and political factors which provided the real impetus to change. The shortage of mathematics teachers in schools, and particularly of graduates, was beginning to arouse concern in the late 1950s. Developments in the use of mathematics in industry during and immediately after the war, and the growth of operations research, linear programming, and computing, had caused an increasing flow of mathematics graduates into industrial employment. The supply of mathematicians available to teach in schools was reduced and a vicious circle seemed imminent, in which a shortage of qualified mathematics teachers would inevitably reduce the numbers of children capable of studying mathematics to a high level – children on whom the future supply of teachers depended. It was almost self-evident that curriculum reform in mathematics would do nothing to solve this problem in the short term; it was less evident that it would not necessarily improve the long-term situation either. However, hopes were high that the new curriculum would make mathematics more attractive, leading to an increased flow of entrants to mathematics courses and ultimately into teaching. In point of fact it is likely that this public alarm was effective more directly, by encouraging an increased intake into mathematics departments in universities and colleges of education. Curriculum development may begin for a variety of reasons, not all of them cogent.

Another immense fillip to curriculum change in mathematics was given when in 1957 the U.S.S.R. launched the world's first artificial satellite, the *Sputnik*. This astonished many politicians and scientists, demonstrating that the U.S.S.R., after many years of supposed technological inferiority, must now be at least the equal of any other major power. It was clear that the U.S.S.R. was devoting a considerable portion of its educational effort towards science and mathematics. In the U.S.A. money was speedily made available to educators who claimed to have the secret, namely curriculum reforms in mathematics and science, and immense programmes were funded. In the U.K. funds on a rather more modest scale were made available by industry to promote developments in mathematics and science teaching – both the Mathematics in Education and Industry Project (M.E.I.) and the School Mathematics Project (S.M.P.) benefited in this way and an Industrial Fund was created to improve the scientific facilities in a number of independent schools. In Europe, too, curriculum change in mathematics was in the air, and some very radical programmes were devised based upon the ideas of the French 'Bourbaki' group of pure mathematicians (who had been influential in reforming the university mathematics curriculum).

Paralleling this ferment of activity at the secondary and tertiary levels were the developments in primary schools. These owed their origin, in the main, to two factors – the gradual abolition of the eleven plus selection process, which freed primary schools from the constraints of a narrow arithmetic syllabus ('mechanical' and 'problems'), together with the spread of ideas, derived from the work of Piaget and other psychologists, about the place of practical experience and activity in children's learning. One of the innovators wrote:

The changes now affecting mathematics are part of a wider movement enveloping the whole pattern of education ... the research undertaken during the past twenty years provides unchallengeable evidence that sound and lasting learning can be achieved only through active participation. (Schools Council 1965, p. xv)

These changes are examined in more detail in chapter 8. At this stage it is enough to summarise by saying that they involved an extension of content to parts of mathematics such as algebra and geometry, which had formerly been the prerogative of the secondary school, and also, more importantly, a change of emphasis in teaching method so that children worked less as a complete class unit with the teacher as the sole source of information. In the new approach, groups of children or individuals followed assignments which might involve movement about the classroom, practical activity, or investigations which the children themselves directed. Some of these developments in teaching method were to have far-reaching consequences and implications when curriculum development in the secondary schools moved into its later phases.

2 Content and method

In the early sixties curriculum reformers strongly urged that much of the content of current mathematics teaching was outmoded. Most of the Schools Mathematics Project texts contain the now famous passage:

This project was founded on the belief, held by a group of practising school teachers, that there are serious shortcomings in the traditional school mathematics syllabus and that there has been for many years a need to bring mathematical curricula into line with modern ideas and applications.

The notorious problems on filling baths from which the plug has been removed, tedious arithmetical calculations of the type used in commerce at the turn of the century, rote-learning of geometrical theorems for reproduction under examination conditions, all came under attack. Dr G. Matthews writes:

Multiply £29 13s 6½d by 37.
Simplify

$$\frac{3x^2 + 5}{x^2 - 5x + 6} - \frac{7x^2 + 4x + 1}{x^2 - 9x + 14}$$

Prove that the base-angles of an isosceles triangle are equal.
These types of question no longer represent mathematics. In fact much of the 'mathematics' traditionally taught in schools is moribund and due for reform. (Sherlock 1964, foreword)

These criticisms were generally justified and arose because the aims which had originally motivated the inclusion of such topics were no longer relevant or had been lost sight of. In commerce the advent of office machinery such as electro-mechanical calculators (soon to be followed by the rapid spread of computerised data-processing) meant that the clerical and arithmetical skills required in an earlier age were no longer a major objective of the education system. The taps–bath situation, a difficult problem when approached *ab initio*, which had become one of a standard set of trick examples taught for exam purposes, is of some interest as an illustration of the idea of rate and also as a mathematical model of a reservoir. (At the present time storage theory is a field of considerable activity, another example of the perils of describing any particular piece of mathematics as 'useless'.) Geometrical theorems merit consideration as examples of the method of logical proof from stated assumptions. The concept of logical inference is a distinctive contribution of mathematics to the curriculum, but geometry was in danger of becoming a meaningless ritual whose only justification was the examination, and reproduction of theorems had been under attack for some time.

What new material was it appropriate to include in the mathematics curriculum? The gap between the old material and the interests of mathematicians both in research and in industrial and commercial applications, plus the advent of the computer, with all that this implied for mathematical processes, suggested a variety of topics. Among these were probability and statistics, functions, relations and sets, vectors and matrices, together with others such as linear programming, group theory, networks (an example of an elementary aspect of topology), a treatment of geometry by transformations or vectors rather than by congruence, and an attempt to link geometrical and algebraic thinking. A new emphasis was given to algebraic structure, and flow-charting and programming with some ideas of numerical analysis at the advanced level were thought to be useful. Some traditional material had to be left out; at the sixth form level this generally implied a reduction in the work on Newtonian mechanics, and the omission of statics, these being replaced by probability and statistics, perhaps with numerical analysis. At all levels there was a reduction in the requirement for complicated manipulation. Some of the ideas were not particularly new; the 'Jeffrey syllabus' of 1944 (Cambridge Local Examinations Syndicate) had emphasised the importance of the function concept, and the inclusion of three-dimensional work in geometry had been advocated for many years. Now some of these topics were taken up.

In addition to all this new content there was to be a particular emphasis on learning with understanding, rather than learning by rote but this, not surprisingly, proved easier to desire than to achieve. To be fair, the curriculum builders were anxious not merely to replace one set of subject content by another but were also concerned with the way in which this material would be taught. Consequently 'drill and practice' examples were greatly reduced in number and many of the exercises involved pupils in exploratory work. In the event these often turned out to be rather difficult, and the lack of straightforward routine examples (which might be given for homework, for example) was considered a stumbling block by many teachers using the materials. The stress on understanding as opposed to rote learning is one which is closely related to the concern with 'process' (considered later in this chapter).

The new emphasis on structure in mathematics was more restrained in the U.K. than it was on the Continent or in the U.S.A., where more radical reforms were advocated. Certain topics were considered to be 'important in mathematics' – thus ideas such as commutative and distributive laws were introduced (leading to the jibe that children now knew that 6×7 was equal to 7×6 because 'multiplication is commutative', without knowing that either product was equal to 42!). An underlying idea here is that mathematics achieves economy by abstraction and generalisation (cf. Bruner's (1960) view that the 'major themes' of a field of knowledge are desirable curriculum content because of their wide explanatory power). An abstract view of mathematics does have a unifying effect, enabling apparently unrelated areas to be subsumed in one general theory or explanation, but this overall view may be helpful only to those who have already encountered the separate instances and are thus able to see the picture as a whole. Arguably, to start at the top and work down is *not* the best way of learning mathematics.

Implicit in the idea of 'important mathematics' is the question – important to whom? The question – what is central to mathematics? – receives no unanimous answer from mathematicians. Generally the pure mathematicians have imposed their view in most curriculum reform, though not without spirited protest from the applied mathematicians (Hammersley 1968; Heading 1971; Dunning-Davies 1975).

An approach to curriculum planning, often markedly unsuccessful and happily generally avoided in the U.K., is to set up a joint committee of educators and professional mathematicians. The hope is that the professional mathematicians will take care of the mathematics, and the educators of the teaching. Unless there is a high level of mutual comprehension – and plenty of time for discussion – such a collaboration is not easy, and in some instances disastrous consequences ensue. (See Polya 1965, p. 134, for a delightful anecdote.) A notorious instance of the adoption of half-understood ideas derived from the priorities of professional pure mathematicians was that of the topic of 'sets' – perhaps nowhere else was the superficiality of a mere change of content more evident (see Freudenthal 1973, chapter 15). At one period, 'sets, and that sort of thing' seemed to be the layman's encapsulation of 'modern maths' – and one shared apparently by many teachers. Hardly a meeting went by but someone would assert that 'sets must permeate all our teaching' or 'sets are the foundation of mathematics'. Why was this?

In some instances the concepts (of the theory of sets) permit the clarification as well as the simplification of the mathematical vocabulary of the students. On other occasions they permit the statement of a mathematical property by means of a simple expression or a brief formula. (O.E.E.C. 1961, p. 13)

Some point out that mathematics concerns itself with sets of objects, relations between them, and operations upon them. If children are introduced to the notation used by professional mathematicians they will have nothing to unlearn as they progress through the educational system. Others argue that the question 'solve the equation $x^2 = 4$' is unanswerable – because we have not stated the permissible set from which x may be chosen. Yet our chief concern is whether the material and notation are appropriate for the children we teach. Logical foundations are not the same as psychological ones; in teaching we are not (solely, or often, or even ever?) concerned with establishing foundations, and arguably the proper place for an examination of the mathematical basis of a course is at the end of it rather than the beginning.

Mathematical activity

A feature of mathematics which recurs continually in any discussion of teaching and learning is that it necessarily involves active participation. This is not generally realised; many laymen would probably expect that a piece of mathematics, properly described, should be intelligible to the reader in the same way that one expects English prose to be understandable. A few mathematicians – very few! – have the ability to communicate complex ideas lucidly to the non-specialist, but even the best communicators of mathematics are powerless to convey mathematical ideas to a passive audience. It has been succinctly put: 'Mathematics is not a spectator sport'.

In each volume of the New Mathematical Library (an enrichment series of background books for teachers and able students produced by the American School Mathematics Study Group) appears a 'Note to the reader', part of which reads as follows:

If the reader has so far encountered mathematics only in classroom work, he should keep in mind that a book on mathematics cannot be read quickly. Nor must he expect to understand all parts of the book on first reading . . . The best way to learn mathematics is to do mathematics, and each book includes problems, some of which may require considerable thought. The reader is urged to acquire the habit of reading with paper and pencil in hand; in this way mathematics will become increasingly meaningful to him.

The amount of detail which any reader needs to fill in will obviously depend on his background – so that to attempt to make an argument clear to every reader may produce a version which is impossibly long.

But there is a more important sense in which the attention of the reader must be engaged, for he must follow the argument in detail as it unfolds. One may watch some television progammes – travelogues or circus spectaculars – with half-an-eye, but detective thrillers demand constant attention to what is happening and to the implications of what is seen. So it is with a mathematical explanation – the reader or listener must be involved.

This aspect of mathematics relates to its twofold nature – for not only is mathematics a body of technique, methods, and results, it is also a 'process' – a method of tackling problems. It may be useful to know that a quadratic equation may be solved by the use of the formula $x = (-B \pm \sqrt{B^2 - 4AC})/2A$, but a mathematician is not one who knows this formula (which, with other similar formulae, can be looked up in books of reference), but one on whom we can rely to solve problems – the criterion is not one of knowledge alone but of an operational ability. Branfield (1969) brings out this point nicely.

A more detailed discussion of the ideas underlying the theme of 'mathematical activity' and some developments which have arisen from them follows in chapter 13; for the moment only a few simple illustrations of the implications for teaching are considered.

In a contribution to Association of Teachers of Mathematics 1966, C. T. Daltry quotes a student: 'Prof. T. was the best lecturer we had; he always seemed to go from the problem to the mathematics. All the others went the other way round.' In this spirit many teachers have sought 'situations' or 'investigations' from which significant mathematical ideas can be developed, many of which have been reported in *Mathematics Teaching*.

As another simple illustration, consider figure 2.1, a regular pentagon in which all the diagonals are drawn. One may say either: (a) 'What do you notice about AB and EC?' or (b) 'Prove that EC is parallel to AB.' In (a) pupils are encouraged to think for themselves what might be the case – a situation much more akin to real-life applications than is (b). When groups of first-year secondary children were given the diagram with the injunction 'For homework find out as much as you can – give proofs where possible,' the results were very interesting, ranging from the (perhaps inevitable) 'I couldn't

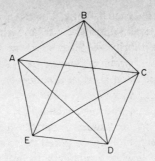

Figure 2.1

find anything, sir!' (a short homework, that!) to two-and-a-half sides and mild remonstrance from parents that 'He spent half the evening on it and we had to chase him off to bed.'

There is nothing new, either, in introducing some topics in such a way that pupils think they have 'discovered' them (as, indeed, in a 'local' sense, they have). There is more interest and more commitment this way. Thus pupils can draw and measure the angles x, y, and z, and discover the result $x = y = \frac{1}{2}z$ (see figure 2.2). 'When you have satisfied yourself that the theorem is true, you start proving it' (Polya 1954).

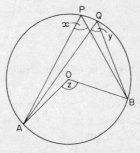

Figure 2.2

It may be objected that many children will discover that $z = 120°$, $y = 59°$, and $x = 61°$ – or other perverse near-misses; it may then be best to leave the situation open – let the class decide whether x and y

ought to be equal, and each half of z. Too often, the teacher's verdict is accepted as the final authority — but in mathematics pupils need not rely on the authority of teacher or the book. (Our science colleagues face similar problems when a bench demonstration or the childrens' own experimental work produces 'wrong' answers. The temptation is strong to say 'Never mind what we actually saw — we should have found that $p \times v$ is constant!' 'Experimental error' has been cynically defined as 'failure to get the result specified by the textbook' — a phenomenon encountered years ago by Galileo in his work on falling bodies when he noted that, contrary to Aristotle's 'Introduction to Physics', heavy bodies do *not* fall faster than light ones!)

Some mathematical educators would wish to draw a distinction between the very limited situations illustrated here, in which the single end-product was of the teacher's choosing, and more open-ended investigations in which the pupil may attend to whatever features of the situation he finds interesting. Many such 'starting points' are described in a book of that title written by three members of A.T.M.

... such situations should be able to be developed in many directions ... they must be presented with flexible and experimental intent, often with deliberate ambiguity ... part of the activity is the formulation of problems that may arise out of the definitions and rules that are developed in discussion of the situation. Students will readily wish to solve problems that they have invented themselves ... (Banwell and others 1972, p. 67)

Situations which they suggest (with possible developments) include:

1 Draw two circles and mark their centres. (Relative positions, tangency, equal-sized circles — join points of intersection to centres, obtaining a rhombus ...)
2 What can you say about $5 \times 5 \times 5 \times 5 \times 5 \times 5$? (Magnitude, last digit — patterns of last digits, e.g. $7 \rightarrow (4)9 \rightarrow (34)3 \rightarrow (230)1 \rightarrow \ldots$)
3 Fold a piece of paper in half. Now fold it in half again. (Powers of 2, sequence of 'up' and 'down' folds, pagination of a book ...)

This aspect of teaching is currently far too much neglected, and, even if one does not completely share the view of mathematics teaching suggested by these writers, one can (and indeed must?) nevertheless make use of problems in the approach to more conventional topics.

There is a sense in which every good mathematics lesson incorporates a problem. For teaching mathematics is not simply imparting information; it is inculcating a way of looking at things, invoking a spirit of enquiry. There should be in every lesson something to be discovered; how are we to do this? – why does this pattern appear? – when will your guess be true and when false? (H. M. Cundy in Association of Teachers of Mathematics 1966, p. 27)

This approach through problem-solving is also strongly supported by G. Polya, himself an outstanding research mathematician whose series of books should be obligatory reading for all teachers and advanced students of mathematics.

What the teacher says in the classroom is not unimportant, but what the students think is a thousand times more important. The idea should be born in the students' mind and the teacher should act only as midwife . . . Let the students discover by themselves as much as feasible under the given circumstances. Much more is feasible than is usually done, I am sure. Let me recommend to you just one little practical trick: let the students actively contribute to the formulation of the problem that they have to solve afterwards. If the students have had a share in proposing the problem, they will work at it more actively afterwards. In fact in the work of the scientist formulating the problems may be the better part of a discovery, the solution often needs less insight and originality than the formulation. Thus letting your students have a share in the formulation you not only activate them to work harder, but you teach them a desirable attitude of mind. (Polya 1965, p. 105)

Polya gives many examples which not only illustrate his ideas on teaching but almost invariably illuminate the mathematical content of the topics he discusses.

Enough has been said to indicate the importance of this aspect of work in mathematics. Unlike change of subject content, however, any change of emphasis in this area is a much more difficult matter, depending as it does on the way in which the teacher approaches his task in his own classroom, beset by a multitude of constraints, many of which press him toward a careful and efficient exposition of the work to a passive class.

3 The pupils

The spectrum of ability

Most curricular reform of the early sixties in Britain concentrated on the programme for able pupils. This was so in the sciences as well as in mathematics – the early Nuffield science projects were for O level biology, chemistry, and physics. Although the Midland Mathematics Experiment, initially based on a group of schools in the Birmingham area, did make a conscious attempt to cater for pupils in technical schools as well, the focus of attention of most of the early schemes was the grammar school (or public school) O level candidate. The best known of these is the School Mathematics Project (S.M.P.), which originated in 1962 in a group of eight schools (Battersea Grammar, Charterhouse, Exeter, Holloway Comprehensive, Marlborough, Sherborne, Winchester, and Winchester Girls' High School). S.M.P. initially produced a two-year O level course, corresponding to entry to the independent schools at age thirteen. Originally grafted on to a series of textbooks (Mansfield and Thompson 1962) appearing at that time, this course was soon replaced, as more maintained schools joined in the activity of S.M.P., by a complete O level course for the eleven to sixteen age range. The S.M.P. 'statement' (see p. 9) indicates the initial emphasis. With their new content and paucity of routine examples, these texts, though by no means ill-suited for use in the participating schools, were generally considered rather too difficult for any but the most able pupils.

Teachers in comprehensive or secondary modern schools, finding nothing designed specifically for their needs, tried out the O level materials with their less academic pupils, easing the pace somewhat and adapting the material to make it more palatable. After a period of use, and particularly as it was felt that the original S.M.P. texts (books 1–4) were rather difficult even for some pupils who would reach O level standard, a revised S.M.P. series was produced (books A–H). This was aimed primarily at the C.S.E. candidate, with a group of transition books (books X, Y, Z) to extend the work for those who wished to proceed to O level. Later still came a further simplification and adaptation of the material of the S.M.P. 'letter' books to produce a set of work cards.

The spread of comprehensive re-organisation and the promised raising of the school leaving age helped to focus attention on the requirements, largely unmet, of the less able. Gradually the size and difficulty of the problem began to be realised, but not until the setting up of the Mathematics for the Majority Project and its offspring the Continuation Project (see chapters 10 and 11) was any large-scale action taken.

Although it antedates the wave of 'reform' and its suggestions were not 'modern' by the criteria of the sixties, the Mathematical Association's Report, *Mathematics in Secondary Modern Schools* was a first step:

The mathematical ideas and teaching methods here discussed are those which we consider suitable for the great bulk of the secondary school population between 11 and 15 years of age, who have shown no early signs of readiness for mathematics as an abstract study, or at least have not achieved the attainments in arithmetic traditionally associated with the ablest of their age group. (Mathematical Association 1959, p. 1)

This quotation highlights the magnitude of the problem – very little provision had been made for the 'great bulk of the secondary school population between 11 and 15 years of age.' Though during the preceding thirty-six years the Mathematical Association had published more than a dozen reports dealing with particular aspects of teaching mathematics to the able pupil, this report was the first to deal with work for the 'other' pupils.

Changing patterns of work in business and industry meant that earlier ideas of what was appropriate would need modifying.

Traditional courses in English schools were planned for a nation of shopkeepers and applications to commerce and trade formed a very large part of them ... We ask that the traditional mathematics syllabus be reviewed very critically and the purpose for which any of it is included be reviewed. (ibid, p. 3)

Nowadays we might want to add that the teaching approaches and indeed the whole aim of the course, not only the syllabus, should be subjected to constant scrutiny, though as chapter 11 will show, we are still far from clear about what this aim should be.

There was a danger that the mathematics curriculum of the secondary modern school would be seen only as a watered down version of O level – the importance of relating the work in a realistic way to the requirements of the pupils is a point stressed in the opening pages of the report:

We hold that it is essential that modern school pupils should see mathematics as a subject that touches their lives and is worth their attention ... There will be relatively few who develop in later years to the stage of taking any University course or of entering professional employment ... (but) there will be those for whom particular parts of mathematics will become an important tool in their chosen vocations ... (ibid., p. 2)

Rapid and continuing changes ... the growing importance of mathematics as a language for the description of new experiences, lead us to suggest that there must be new emphasis in choosing what is suitable material to present to pupils at all stages, and that adaptability and understanding in whatever range of work is undertaken, are qualities to be prized ... (ibid., p. 6)

It is interesting to observe the emphasis given here to practical work, and to work by pupils in pairs or in groups of six to eight, which followed the suggestions made in the earlier primary report (see chapter 8). Important, too, is the weight given to the pupils' need for experience:

We feel strongly that all pupils need actual experience and experiment again and again as they make advance in mathematical expression and procedure. Verbal description of a situation is not enough and mere learning of rules will fail to carry over into new contexts (ibid., p. 19)

The report does not favour the introduction of external examinations for such pupils:

... the danger is that such examinations would overstress the easily examinable elements of courses and so tend to distort the courses themselves and to prevent them from serving the needs of individual schools ... It is not easy even in examinations internal to the school to give various elements of a course a balanced assessment. (ibid., p. 73)

This recommendation was soon to be overtaken by the introduction of the C.S.E. (The dangers pointed out are still there, however, and though some excellent work has been done under Mode III conditions, problems remain.)

By this stage many pupils have already demonstrated an aversion to and lack of aptitude for abstract thought and their motivation in mathematics is particularly low. Though the investigational approach referred to in the preceding chapter may appear unpromising in such circumstances, in the hands of sensitive teachers, approaching their task with realism, it has sometimes been found surprisingly successful.

Seventeen years and two major projects later (see chapters 10 and 11) the problems, and possibly some of the solutions, are somewhat clearer — but what sort of mathematical diet is appropriate for 'the great bulk of the secondary school population' is still largely an unresolved question.

Individualised instruction

As soon as one considers the problems of teaching a wide ability range it becomes apparent that the differences in children's rates of learning and their different motivations and interests must be taken into account. This is part of a general concern not restricted to mathematics, that educational provision should be appropriate to (and perhaps even determined by) the interests and abilities of the individual child, rather than that he should be treated as a unit to be processed through the system. Though many of these ideas of child-centred education originated in the primary school, they were soon under discussion in secondary schools as well, as part of a general re-thinking resulting from problems of re-organisation. Other pressures towards the individualisation of work in mathematics came from teachers' attitudes to the subject itself, such as the ideas of 'mathematical investigation' and 'activity' discussed in the previous chapter. The whole intention of this work was that children should pursue independent lines of investigation, and it was quite impracticable to require that all should simultaneously investigate the same topic in precisely the same way. Other forms of activity likewise presupposed that different individuals would be doing different things at the same time. This was so in many practical activities such as measuring the school playground, carrying out a traffic census, and using scarce equipment (such as the desk calculators at that time finding their way into schools).

Older pupils prefer to be treated as individuals, and a publication of the Mathematics for the Majority Project, *Assignment Systems* (p. 82), strongly supported the use of individual assignments (in the form of conventional work cards or other equivalents), providing examples of work cards and giving suggestions for their use. These are also helpful in dealing with problems caused by absence.

The work card version of the S.M.P. 'letter' course has already been mentioned, and other commercially produced sets of work cards are now becoming available. The B.B.C. television series 'Maths Today' had advocated the use of teacher-made work cards and the B.B.C. later produced an extensive set of such cards to support pupil activities arising from the television programmes. Even earlier, Dienes' Multi-base Arithmetic Blocks and Algebraic Experience Material were each based upon work card systems. Indeed Dienes had said:

If there are different ways of understanding the same piece of mathematics which vary greatly from child to child, then streaming will not succeed in making the situation homogeneous and serious bottlenecks will still occur

. . . It would probably be necessary to abolish almost completely the present method of class teaching with the teacher pontificating from a central position of power, and to replace this by individual learning or learning in small groups from concrete material and written instructions, with the teacher acting as guide and counsellor. (Dienes 1960, pp. 19, 29)

Streaming was an attempt to produce more homogeneous groups to enable the systematic coverage of material by the whole class working together. Moves towards the teaching of children in mixed ability groups, mostly originating from teachers in other curriculum areas, have re-inforced pressures towards the individualisation of instruction in mathematics, for to teach mixed ability groups as a whole class becomes well nigh impossible. The changed mode of working is viewed by some mathematics teachers with mixed feelings. Are the social advantages obtained nullified in the case of mathematics by the difficulties and strains imposed on the teacher attempting to cope with the new situation? Many teachers have had little contact with these methods – there is no well-established body of experience and indeed the effectiveness of mixed ability teaching in mathematics in general is not yet clear.

Besides permitting individual activities, work cards can also foster cooperation between children, encourage self-reliance and independence of the teacher and enhance children's skills in extracting information and getting on with the task. They allow better utilisation of scarce equipment and are almost essential for practical work in mathematics and science. They raise a number of problems, too. If the choice of cards is left to the individual child, will he cover the syllabus? The burden of teachers coping with several different lines of thought simultaneously is considerable (the provision of readily available answer sheets can be very useful here). There are difficulties in the production of cards, particularly in the first setting up the system, and also physical problems of maintaining the system, of coding cards, of monitoring and recording progress, etc., and of ensuring that children have adequately covered the ground on each card before they pass to the next one. (Significantly, the S.M.P. cards include instructions which ensure that the child reports to the teacher periodically to discuss his work before proceeding further.) A major problem is that of ensuring that the level of language is appropriate; if the vocabulary is too difficult the card will be unintelligible. Presenting the card in the form of a tape-recording circumvents some of the difficulties – yet is it not important to give children practice in reading instructions, even assuming the availability of playback tape-recorders? Ideally, assignment cards made up by the class teacher are the most useful since they can be expressly designed for the particular group of pupils. There are, however, immense practical difficulties in this, and the usual approach, at least in commencing such work, is to take over an already-available set of cards produced elsewhere. Cooperative production of work cards by groups of teachers, for example at teachers' centres, is an interesting possibility. Workshop meetings at Keele in connection with the Mathematics for the Majority Project indicated that this was a very useful form of teacher activity and, as we shall see in chapter 11, this was an important aspect of the Continuation Project.

A decade ago there was considerable interest in the production of programmed texts. The ambitious hopes of the early days were rather over-optimistic and programmed learning is no longer considered a panacea. Good programmes turned out to be more difficult to write than had been imagined, but where appropriate programmes *are* available they can be very useful in providing individual instruction (e.g. for a child who, perhaps through absence, needs to catch up on a particular topic which has been covered by the rest of the class). Some of the early workers in this field have shifted their attention to computer-assisted instruction. This has not been widely attempted in this country, though considerable amounts of money have been spent in this area in the U.S.A. (see Hooper (1971), pp. 411–23). Computer-assisted instruction is inherently much more expensive than traditional methods, at least at present. Generally the interest here has been on research rather than practical development. (An ambitious scheme for computer-assisted mathematics instruction in Glasgow schools proposed within the last few years was shelved on grounds of expense). Work at the University of Leeds on elementary arithmetic has been as much concerned to investigate the way in which children learn as to develop ways of teaching. There have also been attempts to let the computer take over some of the management tasks involved in individualised instruction, such as recording pupil progress, suggesting the appropriate next assignment (and providing a copy of this), providing and marking appropriate sets of post-test questions, and so on. An example is the Hertfordshire Computer Managed Mathematics Project. This is 'a computer-managed system for teaching mixed-ability mathematics in the first two years of the comprehensive school'. Over the four years of the project (1973–7) it is planned to increase the numbers from eight schools involving 1,200 pupils (1975) to twelve schools and 4,000 pupils.

A new mathematics course is being developed with the emphasis on individualised work sheets in addition to live teaching and the use of videotaped materials. The computer marks tests and prescribes which work sheets the child should undertake next. Two schools are using the system with on-line terminals and the rest work in batch mode with a courier service (Hooper 1975, p. 95).

An interesting experiment in the use of work cards for teaching mixed ability groups was initiated in nineteen schools in Fifeshire by G. Giles of the University of Stirling (Crawford 1975). Giles was approached by local advisers and teachers who were worried about the consequences of introducing mixed ability teaching for mathematics into the first year of Scottish secondary schools. A series of experimental work cards was devised; their effects in the classroom and some of the considerations used in devising and modifying them are explained in detail in chapter 2 of Crawford's book. Difficulties caused by the sequential nature of mathematics gave rise to doubts among some of the Fife teachers in the initial stages of their work. At first children were given freedom to choose any card, and to work through the cards in any order. This was later modified by combining cards in booklets of eight, all on one theme, and suggesting an order of development through the booklets (though as there were many parallel routes, a considerable measure of freedom of choice remained). Some of the teachers spent only part of the time on work with the cards, the remainder being devoted to class-teaching; indeed in the schools where individual work was used exclusively at first, reversion to a mixed pattern of individual and class work eventually took place.

An important feature of such individualised work is the need for discussion; among the children themselves on the work they are doing and, in particular, discussion with the teacher, who alone may be in a position, by question or suggestion, to bring out the mathematics inherent in the task being undertaken. It helps the teacher, too, in gauging pupil's progress and diagnosing difficulties. But there will be many instances where, to avoid the same point being made repeatedly with groups of children, it may be useful to have a more general class discussion on a particular topic, and even the most enthusiastic advocates of individual work have generally emphasised the value of a review lesson in which the whole class discusses with the teacher some of the work which has just been completed. This is a useful way of consolidating knowledge.

Evidently the role of the teacher in this situation is very different from the usual one:

The whole nature of the project required the teacher to adopt a different role in the classroom, that of a classroom manager. The teacher has no explicit control over what the individual pupils are doing. Pupils will be working from any one of approximately a dozen booklets so it is essential that teachers are fully aware of the contents of all the booklets. (Crawford 1975, p. 57)

There are other organisational problems too, in particular those involved in the storage of materials and in the gradual training of groups of children to adapt to working in this way (although many of them have experienced similar ways of working in the primary school). The role of the teacher is discussed in a stimulating article by Wheeler (1970). Provocative and at times even inconsistent, it is nevertheless full of interest. Wheeler suggests that the teacher who approaches his job 'scientifically' will 'consciously withdraw as much of himself as possible so that he will not be interfering with the activity he wants to promote. He must use every means he can find to focus the attention of the children on the problem, which implies that he must efface himself from their attention. But this does not mean leaving the children to their own devices. 'If we watch this teacher at work we see that he teaches the whole class or a group of the whole class much of the time.' Later he says:

'It isn't easy to describe the way in which this teacher works . . . I hope that you will see that it is quite different from the teacher's way in conventional classrooms on the one hand and from the teacher's way in liberal progressive classrooms on the other . . . Contrary to belief, the progressive classroom *delays* the arrival of autonomy in the child because the teacher is never sufficiently in control of the situation to be able to throw him to the edge of his resources at the moment when it is necessary. (Wheeler 1970)

Here, it seems to me, is the nub of the problem. One may wonder in passing how many children in the traditional teacher-centred classroom are 'thrown to the edge of their resources' during the lesson and, with Wheeler, question whether this happens nearly often enough in the individualised-work situation. Yet this, surely, is the prime function of the teacher. There are moments when guidance, a question, the right prompt, will just push the child over the edge, as it were, into some new insight or discovery. The contribution of the teacher is to provide this, if he is to be more than just a storekeeper for the quantities of materials and equipment which are housed in the 'new classroom'. If the children are all working as individuals or in pairs, he will find it impossible to spend much time with each small group, and the opportunity for him to intervene in their thinking is correspondingly reduced. There was, as we realise, much concealed inattention and inactivity in the traditional classroom, even when all the children were apparently watching the teacher in his demonstration, explanation, or questioning. Individualised methods may only make some of this inactivity, or useless activity, the more apparent, but the fundamental dilemma remains; how can the teacher effectively distribute the resource of experience and stimulus which he, and he alone, can provide, so that it is made maximally available to his pupils?

4 Psychological insights

'Nothing new under the sun' is often surprisingly true in mathematics teaching; some of the up-to-date ideas which are enthusiastically expounded by teacher-trainers have a long history, and the research work of educational psychologists may seem to tell us little which has not been discovered intuitively by able teachers:

Make a parallelogram with strips of wood or cardboard hinged at the corners; use elastic for the diagonals. As the figure changes shape, trace the change in lengths of the diagonals. Are they ever equal? What are their maximum and minimum lengths?

1966 or 1956? These words, written by a teacher who had never encountered Piaget's distinction between the stages of 'concrete' and 'formal' operations, are from Siddons and Hughes (1926, p. 166). They antedate by many years the current view of the importance of concrete experience and imply a 'dynamic' view of geometry reminiscent of the early days of A.T.M. (see *Mathematics Teaching*, **24**, (1963), 75–93). Yet before we cynically discount the work of psychological researchers it may be worth pausing to note that other equally influential figures of the past have made pronouncements with which we might not so readily agree. Take, for instance, the views of John Locke on 'moulding the character of the child', or the dictum of G. H. Hardy which casts the pupil in a purely receptive role:

In mathematics there is only one thing of primary importance, that a teacher should make an honest attempt to understand the subject he teaches as well as he can, and should expound the truth to his pupils to the limits of their patience and capacity. (Ministry of Education 1958, p. 155)

The insights from psychology have tended to affect curriculum reform (where they do so at all) at second hand, sometimes merely legitimating ideas which are already current. They have been more influential in the area of primary education (chapter 8). Particularly important has been the work of Piaget and his collaborators who, over many years of patient work, have attempted to chart the development of children's capacity to understand; the growth of such fundamental concepts as those of number, length, quantity, and weight; the development of the ability to 'conserve' (e.g. in perceiving the invariance of the cardinal ('numerosity') of a set of objects irrespective of their arrangement) and to reverse operations; as well as the onset of abstract thought and the capacity to reason logically. The outcome, if this may be summarised in so few words, has been to emphasise the complexity of children's development, the slow rate at which their thinking can be expected to progress, at least initially, the futility of attempting abstract work at too early a stage, and the desirability of allowing children to develop their own mental structures on the basis of experience.

These ideas have been embodied in the work of Stern (1953) and others (advocating the use of structural materials for number work) and that of Dienes whose Multi-base Arithmetic Blocks, Algebraic Experience Material, Logic Blocks (and more recently other 'concrete' materials), are described and explained in a range of books resulting from his experimental teaching (e.g. Dienes 1960).

Some of Dienes' early work was done in Leicestershire, where later Professor R. R. Skemp, developed a series of texts for the Psychology and Mathematics Project (Skemp 1964–8). A mathematician turned psychologist, Skemp was one of the first to attempt to bring the principles of educational psychology to bear explicitly on the problems of teaching mathematics. Earlier textbook writers had relied, generally with reasonable success, on their intuitive ideas of good teaching, born of years of experience in the classroom. Skemp, whose ideas are very readably propounded in *The Psychology of Learning Mathematics* (1971), attempted a detailed conceptual analysis of the inter-relations between mathematical topics (some examples of the analysis are given in chapter 16 of his book). The results are useful in determining an order in which the topics may best be encountered by the pupil – since, for example, he is unlikely to be able to reach an adequate understanding of the sine ratio without some prior grasp of the idea of similarity, of proportion, and of angle. Though many of Skemp's conclusions had been arrived at empirically in the accumulated experience of teachers, one result of such work was the realisation that some topics (such as fractions, directed numbers, and ratio and proportion) are more difficult than might be imagined. Readers are referred to Skemp's book for a fuller discussion; though there seems little prospect of producing a uniquely determined teaching order based on some idealised conceptual map, the approach does seem a fruitful way of attacking problems of this nature, especially in so far as hypotheses put forward may be subjected to experimental test.

Other workers have attempted to analyse published curriculum materials in science and mathematics, relating their progression to the degree of logical complexity and abstract thinking required at each stage (Malpas 1974). Though such analyses have been made *post hoc* – and therefore serve only to 'explain' why pupils have found particular topics difficult – there is no reason in principle why they could not be used at an earlier stage in planning curricula. In any event, if theoretical principles can be developed upon which, in due course, we can base judgements on curriculum materials, this seems likely to be preferable to the hunch and guesswork approach on which curriculum workers have relied to date. Even the partial progress so far made is probably enough to cast serious doubt on the psychological appropriateness, whatever their mathematical provenance, of the more abstract mathematics curriculum proposals which have been put forward in some other countries.

No discussion of 'psychological insights' can omit mention of the 'learning by discovery' issue, in which the most famous contributor has been Professor J. S. Bruner, who wrote:

For whether one speaks to mathematicians or physicists or historians, one encounters repeatedly an expression of faith in the powerful effects that come from permitting the student to put things together for himself, to be his own discoverer. (Bruner 1961, p. 22)

Though the whole concept of learning by discovery has been the subject of intense debate, and though the experimental studies attempting to demonstrate the superiority of one or the other method have been generally inconclusive (where they were not so badly designed as to be almost worthless), the idea has had a profound effect on many proposals for curriculum reform. Much of the early Nuffield science was built upon this foundation, and it has been a credo of other later projects in both mathematics and science. Though the 'discovery' approach was not Bruner's invention, his eloquent espousal of it – particularly in *The Process of Education* (Bruner 1960) – strengthened its theoretical foundations and was influential in disseminating it. 'Bruner is not the discoverer of discovery; he is its prophet' (Shulman 1970 – an interesting review article).

The issue appears still to be one of belief – of a choice of approach to mathematics teaching which is conditioned by one's view of what teaching is and what mathematics is. Some element of 'discovery' by pupils is an essential in the philosophy of Polya and of the proponents of 'mathematical activity' (see chapters 2 and 13).

A body of knowledge, enshrined in a university faculty and embodied in a series of authoritative volumes, is the result of much prior intellectual activity. To instruct someone in these disciplines is not a matter of getting him to commit results in mind. Rather, it is to teach him to participate in the process that makes possible the establishment of knowledge. We teach a subject not to produce little living libraries on that subject, but rather to get a student to think mathematically for himself, to consider matters as a historian does, to take part in the process of knowledge-getting. Knowing is a process, not a product. (Bruner 1966, p. 72)

From his work in the psychology of learning, Bruner suggested that ideas might be represented in the understanding of the learner on three levels: enactive, where materials are manipulated directly;

ikonic, in which mental images of objects are formed; and symbolic, where the images are replaced by symbols which can be manipulated. In some work with Dienes, Bruner and Kenney (1965) describe how children are led to discover the pattern of results which we would express as:

$$x^2 + 2x + 1 = (x + 1)(x + 1)$$
$$x^2 + 4x + 4 = (x + 2)(x + 2)$$
$$x^2 + 6x + 9 = (x + 3)(x + 3)$$

by working first with wooden squares and strips (Dienes' Algebraic Experience Materials), and later recording symbolically what they have done in some such way as $x^\square + x + 1 = (x + 1)^\square$. 'Virtually everything has a referent which can be pointed to with a finger.' At this stage the representation is still closely tied to the particular concrete embodiment (wooden shapes) which was used. Full symbolic manipulative power is only reached when it is realised that the same symbols can be used to represent other embodiments; rings on the hooks of a balance, pets arranged in rectangular arrays, etc. Note the close correspondence with the ideas of Piaget on concrete and formal thinking and the use of structural apparatus in number work already referred to.

Bruner also put forward the view that education in any discipline could best proceed by isolating the basic principles of the discipline and concentrating attention on these. (Arguably this is the only way in which scientists, for instance, are able to cope with an ever increasing body of scientific knowledge — an example would be the way in which ideas such as the periodic classification of the elements subsume a considerable quantity of detailed information about individual elements — and are themselves subsumed in wider explanatory theories such as that of the atomic structure of matter.)

Shulman (1970, p. 44) describes Bruner's position as follows: 'It is asserted that the fundamental principles or structures of disciplines are essentially simple and consequently these simple structures can be taught and learned in an intellectually honest form through any mode of representation.' The second part of the quotation refers to a celebrated statement made in the report of the Woods Hole conference of scientists and educators (Bruner 1960, p. 33): 'We begin with the hypothesis that any subject can be taught effectively in some intellectually honest form to any child at any stage of development.' By misrepresentation, over-simplification, and tearing from context this rapidly became transmuted to – 'any concept may be taught (to) a child of any age in some intellectually honest manner, if one is able to find the proper language for expressing the concept', or even to a version of this which omits the qualifying clause. Coupled with statements like 'children are capable of learning concepts at much earlier ages than formerly thought' (Lloyd Scott, quoted in Shulman 1970, p. 24), this appears to lend weight and respectability to some highly dubious attempts to introduce abstract mathematical concepts ('of fundamental importance in mathematics') to young children. This seems completely wrong-headed and contrary to all the implications of the work of the Piaget school on the natural development of children's thinking.

This can hardly be what Bruner intended; the 'hypothesis' with 'concept' substituted for 'subject' is almost self-evidently untrue. How would one communicate the concept of uniform convergence to a two-year old? Further it was a hypothesis, a starting point; what the members of the conference were saying was surely 'Let us forget that (say) anthropology has until now been studied only at the university and consider the effect of introducing such ideas at school level.' To take off the blinkers and attempt to think out a problem in an unusual frame of reference may be a productive activity – to restrict oneself completely to the new viewpoint may only be to exchange one set of blinkers for another.

Though some of Bruner's views may have been misinterpreted by lesser minds, his contribution to the study of human learning has been immense, and his writings have deservedly exercised a profound influence on curriculum change.

5 The use of mathematics

What use is it?

Although mathematics currently takes up a large proportion of the timetable of almost every child in school, it was not always so (Howson 1974) and may not continue thus (Reynolds 1974). Is it self-evident that this amount of mathematics is appropriate for all children? Questions are asked about its 'relevance'. Though this word has been seriously overworked (especially in universities in the late sixties), it is quite legitimate to ask of school mathematics, 'What is it for?' Ormell (see p. 99) has analysed relevance as *horizontal*, supporting work in other curriculum areas (e.g. mathematics which is obviously and visibly relevant in current work in science), and *vertical*, implying deferred reward (material is taught and assimilated now on the understanding that its usefulness will be apparent later). A justification for the place of mathematics within the curriculum is attempted later (chapter 14), but however we describe the purposes of teaching mathematics in school there is always a danger that the grander notions will be lost sight of in the day-to-day business of teaching, so that the shadow is mistaken for the substance.

We must never take the content of the curriculum as specifying the objectives. In a great deal of formal education, the kind that many of us deplore, I think this is exactly what happens. The content of the curriculum, the works of Shakespeare, or the areas of science and mathematics that are to be used as means for achieving very complex and sophisticated educational ends, is often taken as itself providing the educational objectives. The result is that the education concerned becomes directed towards the mastery of a particular content of propositions when the propositions in this particular area of knowledge should be employed as a means of achieving very complex objectives, by way of particular dispositions, types of intellectual skills, various habits of mind, and the like. (Hirst 1968, p. 43)

Thus we may be concerned to promote logical reasoning through a study of geometrical propositions about circles. Here the circles are not important; the logic is.

Any discussion of relevance is intimately connected with the motivation of the pupil, for if he feels that his mathematical work is entirely divorced from reality, from his own life and interests, there is little likelihood that he will achieve success. No universal recipe for motivation has yet been found, but the importance of *success* and of *support* are quite evident. Some aspects of mathematics are extremely difficult at first encounter; it is hard for the teacher, usually an experienced mathematician, to recapture the feeling of confusion and isolation which is often the lot of the tyro. (The benefit for a teacher in solving mathematical problems at his own level is not solely, or often even primarily, because of the improvement in his own mathematical knowledge which results, but because it helps to resensitise him to the problem of his pupils.) It follows that the teacher must not place his pupils in a situation where they are constantly plagued by failure. If they can be given some success in their work and if he is able to support them in the difficulties which they encounter, they are more likely to be motivated to work at their mathematics.

Some have sought to find relevance in mathematics by relating it to the environment, notably in the work of the Mathematics for the Majority Continuation Project (chapter 11). There are difficulties here, however, which we take up later; mathematics is not so self-evidently present in the environment, and some of the mathematics arising from environmental problems is distinctly difficult and complex. However, we should not forget that for a child, the world in which he lives may well include games, puzzles, and man-made situations (which may have been introduced into the child's environment precisely so that they could form the basis of mathematical activities). In this light, the mathematical 'investigations' considered in chapter 2 are not so 'artificial' after all.

A service subject?

Early curriculum reform in mathematics in the U.K. was in general little concerned with its relationships with other curriculum subjects (though some European schemes were even more inward-looking). Early attempts at liaison between the Nuffield Science Projects and the concurrent developments in S.M.P. came to nothing and each of the curriculum projects tended to go its own way, the scientists using mathematics with reluctance only when forced to do so. A similar 'isolationist' attitude lay behind the Physics-with-Mathematics A level examination of the Oxford Local Board. (Here it is envisaged that the mathematics required for the A level physics course will be taught by the physics teacher as an integral part of the course.)

'Some will tell you', wrote Bickley (1958), 'that mathematics is the Queen of the Sciences; others that she is their handmaiden.' The tension within mathematics teaching between the role of mathematics as an independent subject in its own right and its service function to other curriculum areas is recurring and unavoidable. Unilateral declarations of independence on the part of mathematicians or of scientists are not likely to be helpful, though it is not always easy to see how the various conflicting demands can be met. In chapter 12 we consider a project which devoted considerable attention to this difficulty at sixth form level.

One of the priorities is to establish what mathematics is required and to sort out the problems of sequence, of synchronisation, and of overlap. Thus children may need to be able to use tables of sines or of reciprocals for work in physics before they encounter these topics in mathematics. At a later stage in sixth form work there may be problems if the same piece of work on simple harmonic motion is covered twice, once by the mathematician and again by the physicist – perhaps using an entirely different notation (to say nothing of the unseemly haste with which some mathematicians have been known to rush to cover this topic 'before the scientists have time to implant all the wrong ideas'). The root of the problem is usually a lack of communication between the teachers concerned, each busy with other duties.

It seems important, too, to establish the true requirements of the user subjects. Sometimes scientists complain that four-figure logarithms are no longer used. Some of the modern syllabuses advocate the use of three-figure logarithms; (S.M.P. initially even avoided logarithms altogether in favour of the slide rule). Yet does the accuracy achieved in most scientific measurements (other than those of mass) justify the use of more than three significant figures (i.e. slide rule accuracy)? And should not the likely limits of error in an experiment be an important aspect of a science course?

Whereas two or three decades ago the only subjects which used mathematics were physics and to a lesser extent chemistry (with minor applications, such as map projections, in geography), the situation has now changed. Biology makes increasing use of mathematics, particularly in the statistical area, while economics, a developing subject at school level, is now another 'user subject', and the amount and nature of the mathematical work required in geography has undergone considerable change. The demands from the user subjects are often conflicting: physics and chemistry require mainly calculus, geometry, and mechanics (the 'traditional' mathematics); biology and geography have different emphases (probability, statistics, and some new material such as optimisation and network theory). Thus it is hardly possible to formulate a mathematics curriculum to satisfy all customers.

Other difficulties arise, too, related to the degree of complexity required in applications and the understanding of mathematical principles involved. A mathematician may prefer to restrict work in some topics to examples involving only 'easy' numbers or formulae which illustrate the principle, whereas the practical requirements of the user may involve manipulation of quite complicated expressions which will be found difficult unless the pupils have been given practice in such skills. Electronic calculators may bring some benefit here.

On the other hand, mathematicians providing service courses are often accused of making the whole issue far too mathematical and complicated: 'We only want to use the mathematics', say the consumers, 'but you won't just give us a formula, you insist on proving it for us.' Chapter 14 examines some recent attempts to build bridges between mathematics and other subjects.

6 Examinations and assessment

Changes

The changes in examination and assessment since the early fifties have paralleled and been stimulated by those in the mathematics curriculum. In 1951 the General Certificate of Education O and A level replaced the School Certificate and Higher School Certificate examinations. Results in individual subjects were now recorded and there was no requirement for students to sit groups of subjects at the same examination, but apart from this there was no substantial alteration in the content of the examination or in the methods of assessment. The first wave of curriculum reform in science and mathematics faced the problem of examinations by requesting a new syllabus for the examination, a right which few schools had previously exercised. The School Mathematics Project (S.M.P.) put forward its own O level syllabus, the Oxford and Cambridge Board administering the examination on behalf of all the other G.C.E boards. The first S.M.P. O level examination, held in 1964, incorporated a change in examining style as well as of syllabus; one of the papers consisted of multiple choice questions. Other projects (such as the Midland Mathematics Experiment, Mathematics in Education and Industry, and the St Dunstan's experiment) were also successful in making special examination arrangements.

Multiple choice questions were not very familiar at that time in the U.K., although such question papers had been in use for some time in the U.S.A. It is well known that the conventional essay-type examination must sample the syllabus and can only cover a small proportion of it, particularly where a choice of questions is allowed. Moreover it tests the candidate's ability to express himself in English as well as the subject matter of the syllabus. While this may often be desirable, there are occasions when it distracts from the main purpose of testing the candidate's competence in the chosen discipline. Admittedly this is not usually considered a major problem in testing elementary mathematics. Multiple choice questions offer some advantages. They permit a very thorough coverage of the area under consideration, and the demands they make on the candidate's powers of self-expression are considerably reduced. The way they are marked is independent of the individual and perhaps arbitrary decisions of examiners, so that a high degree of mark standardisation can be achieved. Again this is not so pressing a problem in mathematics as it might be, for example, in history. A further consideration is that questions can be pre-tested by their being given to an appropriate group of pupils. Analysis of the results enables the elimination of questions that are in some way unsatisfactory, either because they are too easy or too difficult (*facility*), or because they fail adequately to distinguish between candidates of different abilities (*discrimination*). Thus it is possible to build up a bank of questions of a standardised level of difficulty which can be used in examinations in successive years allowing comparisons to be made. Though the marking of such questions is very simple their composition is time-consuming, so that it is desirable that they be kept confidential.

The introduction of the Certificate of Secondary Education (C.S.E.) also brought about some changes in the assessment of mathematics, not only by extending them to a much wider area of ability, but because in many instances provision was made for the incorporation of course work or project work in the formal assessment. A number of schools (smaller than had been hoped) also took advantage of the Mode III provision (which allowed them to construct and carry out their own examinations on their own syllabus work, with moderation by an external examiner). Though these innovations introduced a number of very difficult problems, there were potential gains which were often considered to outweigh the disadvantages. The Schools Council has produced a large number of publications on the problems of examining, several of which are devoted to mathematics.

Problems

A recurring difficulty, which mathematics examinations share with those of other subjects, is the extent to which questions become stereotyped. Griffiths and Howson express this as follows:

Those readers who are unfamiliar with the British system are reminded that many of the questions to be found in these papers, particularly those which examine traditional material, recur in disguised form year after year. In short, they are stereotyped and accordingly candidates will be well practised in their solution to the unfortunate exclusion of other mathematics and skills. (Griffiths and Howson 1974, p. 344)

Copies of past papers are used quite legitimately by teachers and candidates in order to interpret the terminology of the syllabus. (As an example, for many years A level mechanics syllabuses have included 'Newton's Laws, force, mass, and acceleration' and this short list of topics could account for quite a considerable fraction of the work in mechanics studied at university level.) Evidently a syllabus alone is not an adequate statement of the material to be covered. Nowadays this is usually acknowledged by the provision of sets of specimen questions whenever a new syllabus is produced. Nevertheless, over-use of past examination papers can have a detrimental effect. The following was written some years ago for an A level syllabus revision exercise of one of the G.C.E. boards:

It must be admitted that examinations have a considerable influence on what is taught in schools, so that in the majority of cases, topics which are not examined will not be taught. More importantly, methods of approach which do not easily lend themselves to examination by the normal methods will also fail to be used. Consequently there is a considerable tendency for examination questions to take a routine form, of similar type to those the candidate has already attempted in the course of his normal lessons. This can lead and often does lead to a situation in which much of the time spent on mathematics consists of working through back papers, on the type of questions which are expected.

To what extent is it possible to set questions which demand some investigation by the candidate, some inventiveness, such as the formulation of hypotheses, their verification and proof (where this lies within the power of the candidate)? 'Open-ended' questions of this type might well have some influence on teaching, and one hopes, would lead to a greater emphasis on mathematics as exploration rather than the assimilation of techniques and results in cut-and-dried form. I do not under-estimate the difficulties of this approach. There seem to be two main objections, first that questions of this vaguer type are much more difficult to answer, and candidates prefer something routine which they feel they can do; secondly, that it is not possible to mark questions of this nature as objectively as is usual in mathematics. The first of these objections seems the stronger of the two, for I admit that it is not easy to find questions of a suitable nature. However, I think this might be tried. As to the second objection, this problem is one which has to be faced by examiners of other subjects in the curriculum and I think it would be worth the sacrifice entailed if the cramping effect of examinations could be reduced. (Watson 1966)

An indication of some of the difficulties involved in framing 'non-routine' questions and in persuading preparatory committees of their value is given by the following. The last question on J.M.B. Further Mathematics, Syllabus B 1970, Paper 2, read as follows:

A car ferry leaves a jetty at 10-minute intervals and can carry 2 cars. Cars arrive to board the ferry randomly and independently at an average rate of 12 per hour. Assuming that the number of arrivals in any 10-minute period has a Poisson distribution, and given that no cars are left behind at the first departure of the ferry, find the probabilities that:

  (i) no cars are left behind at the second departure
 (ii) no cars are left behind at the third departure
(iii) just one car is left behind at the third departure

(You may leave each answer in the form of a rational multiple of an inverse power of e.)

An earlier draft version of this question involved a four-car ferry. Candidates were asked to find the probability that some cars would be left behind when the ferry departed at the end of any given 10-minute period (assuming no cars were waiting at the beginning of the period). Next they were asked to find the corresponding probabilities if a larger ferry, holding up to 8 cars, were to be installed (and as a result traffic increased to 24 cars per hour), a 10-minute service being maintained. They were asked to consider the new situation from the point of view of (a) the motorist, and (b) the ferry operator, and compare it with the effect of operating a 5-minute service using two 4-car ferries with a traffic density of 24 cars per hour. (An idea inspired by the former Ballachulish ferry!) In this earlier version the attempt to make the problems more realistic by 'increasing n' involved the use of the normal approximation to the Poisson distribution. The later parts of the draft made the question long and probably too difficult, but were also criticised as 'unanswerable, since the required information is not given in the question'. (There is an issue of principle here. Examination questions are sometimes attacked for their artificiality; they seldom (intentionally!) contain redundant, contradictory, or inadequate information. A candidate might legitimately be asked to estimate quantities not given, or to indicate variables which are relevant, but usually each item of information given in a question is essential to the solution.) Griffiths and Howson (1974, p. 334) examine the difficulties of restructuring examination questions so as to make them more evaluative of insight; they give some specimen A level questions followed by comments and suggested re-wording. Clearly there are many difficulties in producing questions which, while fair to the candidates and not impossibly difficult, test insight and ability in non-stereotyped ways. Among the issues which arise are these: To what extent can an examination on a normal syllabus, to be taken under timed conditions, contain such non-routine questions? Should the assessment of these abilities be taken out of the timed examination situation and be tested in the results of course work or project work undertaken independently? Does the use of untimed examinations offer any way forward? The case studies in chapters 12 and 13 indicate some directions in which progress has been made.

The 'objectives' approach

The notion of specifying the objectives of a particular curriculum unit in precise terms came into prominence with the publication of the *Taxonomy of Educational Objectives* (Bloom and others 1956). The original Bloom classification described five categories of educational objectives which are arranged in an ascending hierarchy, i.e. knowledge, comprehension, application, analysis, and synthesis. Each category was illustrated by examples – thus 'knowledge' would include simple factual recall of items such as formulae, definitions, notations, etc. The next level, 'comprehension', was concerned with the routine use of ideas, such as the ability to translate verbal mathematical material into symbolic statements and vice versa, or to perform routine computations or mathematical processes. 'Application' referred to the use of ideas and material in non-routine situations; 'analysis' to such abilities as those involved in recognising unstated assumptions, distinguishing facts from hypotheses, checking the consistency of hypotheses with given information; and 'synthesis' to the putting together of elements and parts so as to form a whole, organising and presenting ideas, proposing ways of testing an hypothesis, and making mathematical discoveries or generalisations. A sixth taxonomic level was proposed, that of 'evaluation', the making of judgements about the value of ideas, solutions, methods, material, etc. This was placed last since it was regarded as requiring to some extent all the other categories of behaviour, although it was recognised that it might occur at an intermediate stage in the process of thinking or of solving a problem. Later workers have modified these suggestions in an attempt to make them more appropriate to mathematics. Thus, evaluation (such as judging the correctness of a proof by a breakdown of the steps) might well be placed in the category of analysis, and seems indeed to be an integral part of the process of proof itself. Avital and Shettleworth (1968) replace the five levels of the Bloom taxonomy by three categories which they label 'recognition or recall', 'algorithmic thinking or generalisation', and 'open search'. They give illustrative examples showing how objectives in each of these categories can be tested, usually by multiple choice test items. The problems of testing the higher levels of the classification in this way are admittedly difficult, though not impossible, as Avital and Shettleworth demonstrate.

Precise attempts to categorise the level of attainment being tested by any given question are attended by the additional difficulty that so much depends on the background of the learner.

For example, two students solve an algebra problem. One student may be solving it from memory having had the identical problem in class previously, the other student has never met the problem before and must reason out the solution by applying general principles. We can only distinguish between their behaviour as we analyse the relation between the problem and each student's background of experience. (Bloom and others 1956, p. 16)

Despite these difficulties the idea of classifying objectives has influenced the production of examination questions and there have been attempts, particularly in some science syllabuses, to provide (as an integral part of a syllabus specification) a statement of the educational outcomes which the examination is intended to measure and the weight to be given to each of the outcomes (J.M.B. 1970, p. 16). Perhaps an attempt to specify exactly what each question will be measuring and the precise weighting to be given to each of these abilities by the examination as a whole is doomed to failure. (The J.M.B. mathematics panel felt so and adopted a less formal statement.) Nevertheless there is some benefit in a more detailed and explicit statement of the purposes of the examination to replace the rather shorter general statements, usually accompanying syllabuses, which teachers have supplemented by their deductions, based on perusal of the published examination papers. 'Just what is this examination for?' is a question which must repeatedly be asked.

PART II

7 Introduction to the case studies

Any discussion of the reforms which have taken place in Britain in mathematics teaching must take account of the wide diversity and large number of projects which have arisen over the years. It is interesting to speculate on the possible reasons for this. The contrast with the way in which parallel developments occurred in science is considerable. In the early years of change the subject teachers' associations in science were particularly concerned with the new developments, and the early initiatives for change came from the Science Masters' Association and the Association of Women Science Teachers. The Mathematical Association, which was well-established and had published over the years a number of extremely detailed and useful reports on the 'state of the art' in mathematics teaching, was less inclined to promote change and appeared more conservative in outlook. At this point the 'rival' Association of Teachers of Mathematics was in its early stages. (It was formed in 1952, originally as the Association for Teaching Aids in Mathematics.)

The policy of the Mathematical Association was to produce definitive reports giving a consensus of good practice in respect of mathematics teaching; of its very nature this approach was unsuited to the tentative experimental situation in which the early curriculum developments took place. In many cases ideas were being tried out for the first time and no one knew whether they would succeed, nor how much the outcome was a consequence of the enthusiasm of individual teachers. The result was that there was a tendency to wait until the innovations had proved themselves before they were given much publicity. This made it more difficult for teachers to learn what was going on, and the first developments were not instituted in any organised way.

The Association of Teachers of Mathematics was more energetic and its book *Some Lessons in Mathematics* (Fletcher 1964), produced at a 'writing week' in 1962, was an early seminal contribution. In the absence of any 'official' initiative, regional groups of mathematics teachers took the law into their own hands and began to try out their own ideas. (The free-for-all in mathematics curriculum has been compared unfavourably with the tidier situation in science. It has undoubtedly had some unfortunate consequences, as when pupils (or teachers) have changed schools, to name but one. Yet there has been no 'new orthodoxy', and a variety of approaches have received attention so that by a process of clothes-stealing – more tactfully labelled cross-fertilisation? – the better ideas have become widely known.)

The early science projects were funded by the Nuffield Foundation, but this was not the case, initially at least, in mathematics. (An exception was the Nuffield Primary Mathematics Project.) Other developments solicited what funds they could from a variety of sources; some received local authority support, others depended largely upon the enthusiasm of the participating teachers, and a few succeeded in attracting donations from industry. S.M.P., which showed particular financial flair, ultimately became largely self-financing when the output of saleable materials (initially mainly textbooks) was large enough to provide a royalty income to finance further developments, and is now managed by a board of trustees.

The early secondary projects were mainly concerned to achieve a change of content, sometimes without any very clear criteria other than that 'it might be interesting to try to teach this material to children in schools'. As developments proceeded the interest shifted for a variety of reasons. These included the realisation that the new content was not necessarily any more productive of understanding than the old, and the difficulties which were experienced in trying to

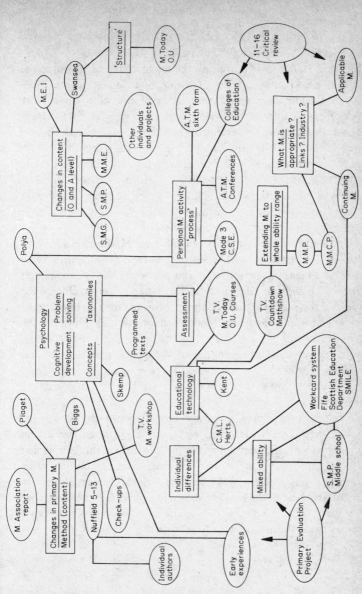

Figure 7.1 Mathematics Curriculum Development in the U.K. (based on an idea of M. Ruth Eagle, 1975). Important issues are underlined. Names of Individuals or projects are ringed. Earlier developments *tend* to be higher on the page, and pupil age-group *tends* to increase from left to right. M = mathematics/mathematical.

teach this or any other mathematical material to the children of lower academic ability increasingly staying on at school.

Figure 7.1 attempts to indicate the major concerns to which the projects were a response. Like all classifications this has its imperfections – no doubt one could argue on the placing of any particular development within the overall scheme, or suggest that certain projects should appear several times in different places on the figure. Some of these developments have been selected for more detailed discussion as case studies in the chapters which follow. There are many candidates for inclusion; the choice is necessarily a reflection of the author's personal interests and idiosyncracies, although an attempt has been made to illustrate the major themes and to give coverage to a range of ages, abilities, and approaches.

A notable omission is a study of the School Mathematics Project. This is documented in one of its own publications (Thwaites 1972) which gives an authoritative statement of its development and rationale (though only from an 'inside' point of view). One of the earliest of the secondary developments, this is now one of the most popular and widespread. It has continued its activities through to the present day and is now redrafting its sixth form material and developing work in the area of the middle school. The 11–16 course has undergone several successive stages of revision; their sequence gives an interesting insight into the changes in approach during the last few years.

Some other developments are also in widespread use, such as the work of the Scottish Mathematics Group. Centralised curriculum planning ensures that developments in Scotland are more rapid and uniform; these texts have also received considerable attention overseas (as have those of S.M.P. and the Nuffield Primary Project, in particular). Many of the projects shown in figure 7.1 are referred to in the subsequent text by way of illustration, and source references are given to assist those who wish to read further.

8 Primary mathematics: the first phase

... children, developing at their own individual rates, learn through their active response to the experiences which come to them; through constructive play, experiment, and discussion children become aware of relationships and develop mental structures which are mathematical in form and are in fact the only sound basis of mathematical techniques. The aim of primary teaching, it is agreed, is the laying of this foundation of mathematical thinking about the numerical and spatial aspects of the objects and activities which children of this age encounter. The justification of this theory and the implications for the day-to-day work of the teacher form the subject matter of this report. (Mathematical Association 1955, p. v)

These words mark the beginning of a widespread change in approach to primary school work in general, and to mathematics in particular. The idea of 'learning through experience', with the consequences for teaching methods which flow from its adoption, is now a commonplace of primary education. It was not so then, and the great changes which have occurred in primary schools, and not least in mathematics, owe their origin very largely to the acceptance of this approach, linked with the freedom of action given to teachers as eleven plus selection was gradually eliminated. The major change was in teaching method, but the content of mathematics in the junior school was also changing.

At first sight the word mathematics may seem pretentious when used of education in the primary school ... Whatever name be used, this report is concerned with an approach to the relations and ideas which are fundamental, not only to the art of reckoning, but also to mathematics as a whole. (Mathematical Association 1955, p. vii)

The report contained suggestions for work on shape, measurement, area, volume, simple graphs, and curve stitching, as well as number. The use of simple apparatus in early number work was advocated and an appendix gave suggestions for activities which could be used to develop children's understanding of the concept of weight. A surge of activity followed in the next few years; courses were arranged by national and local education authorities and by teachers' associations, new textbooks were published, and structural apparatus such as Cuisenaire rods and Dienes' blocks began to appear.

When *Mathematics in Primary Schools* was published (Schools Council 1965), the pattern of these changes was becoming clear. The bulletin was written by a member of H.M. Inspectorate, Miss E. E. Biggs, whose continued and enthusiastic advocacy of the new approach had been an important feature of the developments. It was a 'summary of intensive work in the learning of mathematics by discovery methods carried out with children and teachers' – for the teachers who attended the D.E.S. courses organised by Biggs found themselves learning mathematics by the same 'discovery' or experiential methods as they were urged to use with their pupils:

If teachers are to be convinced that children can learn mathematics through their own discoveries, they must first experience discovery of mathematical concepts for themselves. This experience has been provided, in the first place, by means of courses in mathematics. (Schools Council 1965, p. xv)

The research upon which the new approach was based is discussed in chapter 2 of the bulletin, and the conclusions to be drawn from it are summarised as follows:

1 Children learn mathematical concepts more slowly than we realised. They learn by their own activities
2 Although children think and reason in different ways, they all pass through certain stages depending on their chronological and mental ages and their experience

3 We can accelerate their learning by providing suitable experiences, particularly if we introduce the appropriate language simultaneously
4 Practice is necessary to fix a concept once it has been understood, therefore practice should follow, and not precede, discovery. (ibid., p. 9)

This was the philosophy underlying the 'activity methods' which were now becoming widely used in the primary school. The implications for classroom organisation and the role of the teacher were considerable. In this new pattern children might work as individuals or in groups, as opposed to the whole class working simultaneously on the same task. Practical activity means mobility within (and outside) the classroom, so that grouping of desks or tables in clusters rather than in rows was more appropriate – children were also to be seen counting the number of vehicles passing the school gate or pacing across the playground with a trundle-wheel.

The new approach was not – and is not – a panacea. It is perfectly possible for children to waste time in unprofitable pursuits, or to carry out 'activities' without perceiving the point which lies behind them. Apart from the additional burdens of organisation and control place upon the teacher, who may be required to be in six places at once, he has the problem of guiding and advising children who are engaged in a variety of tasks, like a chess expert engaged in a simultaneous tournament. Such a classroom regime makes heavy demands upon the teacher, especially when it is translated to the secondary field, where the differences between the stages reached by different pupils and the complexities of the tasks they are undertaking are both greater. Nevertheless, the methods have become quite widely used in primary schools right across the curriculum, and in more recent years have been adopted in work with the lower age groups of some secondary schools. Undoubtedly there have been great gains from this approach, but there are also some consequential problems, arising from the 'sequential' nature of mathematics, which we shall examine briefly later.

Another aspect of 'learning by doing' which arose from the increased awareness of work such as that of Piaget, was the introduction of 'structural apparatus' for use in number work. Catherine

Stern (1953) in *Children Discover Arithmetic* parodies the abstract way in which arithmetic is sometimes taught, by describing a 'barbarian land' where the children learn a 'number sequence'.

lul, laa, buy, bay, bee, lol

with some simple addition combinations:

$$
\begin{array}{cc}
\text{lul} & \text{laa} \\
+\text{lul} & +\text{buy} \\
\hline
\text{laa} & \text{bee}
\end{array}
\qquad \ldots
$$

and some subtraction facts:

$$
\begin{array}{c}
\text{lol} \\
-\text{bay} \\
\hline
\text{laa}
\end{array}
\qquad \text{etc.}
$$

('there are, of course, about one hundred such combinations in addition and subtraction which you should learn.')

She also shows how the 'natives' write these examples in symbolic form:

＋ ⅄⁄∠   ＋ ∠⁄▽⁄ℂ   ＋ Ψ⁄∠

Study these combinations and try to remember the answer in each case, calling it by its right name. Cover up the row of answers and see whether you have mastered the examples. Just glue together in your mind the combination ℂ + ⅃ = ⅄ (pronounced 'bee' + 'bay' equals 'taa'). As some psychologists tell us, we have but to repeat and repeat such combinations until the stimulus ℂ + ⅃ brings forth the correct response ⅄ (Stern 1953, pp. 10, 11)

As Stern points out, the number sequences (one, two, three . . .) and the symbols (1, 2, 3) we use to represent them are as meaningless and arbitrary to the young child as are her imaginary 'lul, laa, buy'. Addition and subtraction facts have to be memorised (most people are probably unaware that as well as learning multiplication tables — which they do remember doing — they have had to learn addition and subtraction 'tables'). But if these are presented in an abstract, and therefore meaningless, way to children who are at the age where thinking is related to 'concrete' objects, it is not surprising that children find difficulty in assimilating the ideas. Concrete referents such as buttons and counters have been in use, of course, for many years (fingers are probably the oldest established structural apparatus as well as having advantages of cheapness and ready availability!) and as Stern acknowledges, other systems of teaching arithmetic (Tillich, Montessori, and others) have used blocks. The important thing is to provide concrete instances of which the arithmetical statement is an abstract representation – thus 'the work on paper is merely the recording of a concrete operation'. That is, $4 + 3 = 7$ means something like: 'To a pile (set) of four buttons I added three buttons. When I counted them all there were 7.'

(Stern's 'rods', which are similar in design to several other commercially-produced materials such as Cuisenaire, Colour-Factor, etc., solve an additional problem. A group of 7 buttons looks very like a group of 8 buttons; a child who miscounts has no easy way of realising that his answer is wrong, whereas it is clear that rods of length 3 and 4 units placed end to end are matched by the 7-rod and not the 8.)

Similar considerations underlie the use of the Multi-base Arithmetic Blocks (M.A.B.; see Dienes 1960). Dienes has also produced materials for the teaching of a whole range of aspects of algebra (simple and quadratic equations, indices, using M.A.B, and vectors); logic blocks, or 'attribute materials', for the teaching of

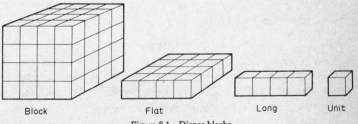

Block   Flat   Long   Unit

Figure 8.1   Dienes blocks

classification and 'set' operations; and other materials for work on relations and functions.

If we are considering arithmetic in base 4, a number of wooden blocks of each of the shapes in figure 8.1 are provided. The child can see that '4 units make a long', '4 longs make a flat', etc., and a 'sum' such as the following:

	Blocks	Flats	Longs	Units
	2	1	2	3
Add	1	1	1	3
	3	3	0	2

is a record of some such activity as: 'I took 2 blocks, 1 flat, 2 longs, and 3 units and John took 1 block, 1 flat, 1 long, and 3 units. When we put our wood together we found we had 3 blocks, 3 flats, and 2 units.' (Note that 4 units have been 'exchanged at the bank' for 1 long, and 4 longs for 1 flat.)

By this means, all the operations of arithmetic which involve place value (addition, subtraction, multiplication, and division of whole numbers and of 'decimals') can be given concrete representation. Dienes advocates that the experience should be varied in two ways:

1 Perceptually, by using other 'embodiments' of the same concept, e.g. with triangles whose sides are in the ratio 1, 2, 4, 8 (and hence whose areas are in ratio 1, 4, 16, 64), or (rather more abstractly) with coloured counters whose exchange values are agreed: 4 white = 1 red, 4 red = 1 black, 4 black = 1 gold, etc.

2 Mathematically, by varying the base (in this case 4) of the arithmetic system, e.g. by repeating the work in base 3 or base 5.

The concept of place value is what these varied representations of different base-systems have in common. It is the underlying concept, rather than only one of its representations, which we wish the pupil to acquire, so that he understands place value in general, and the base-10 system in particular. (It is interesting to observe in passing how this idea has been thoughtlessly incorporated into the 'new orthodoxy' of modern mathematics, particularly at the secondary level. Textbooks have chapters on number bases and in some schools children even practise calculations such as 'Divide 4732_{12} by 35_{12}'. When challenged, this is justified by 'It's in the syllabus/textbook', 'It's modern maths', or 'They need to know about number bases because of computers, don't they?' In fact, though computers perform their arithmetic in the binary system, the user need never know this, since input and output are now invariably in ordinary decimal notation. Computation in other bases can be entertaining and is a test of understanding of place value notation, but it seems totally pointless to overdo this work.)

The aim of using structural materials is to teach certain concepts which are inherent in the use of the materials, in contrast to the less-directed activity of 'finding mathematics in the environment'. There is thus a certain tension inherent in learning by experience, between the freedom which is given to pupils to follow the questions of interest to them, and the teacher's overall aim of introducing the children to mathematical topics which are considered important (and which may be tools for progress in later work).

Biggs, writing in 1972, says:

The chief aims of teaching by discovery or, to use a less controversial term, learning through investigation, are: first to give pupils opportunities to think for themselves; second, to provide experiences from the environment (natural and man-made) which will help them to discover the order, pattern, and relations which form the basis of mathematics; third, to give them the necessary knowledge and subsequent skills. (Biggs 1972, p. 8)

She goes on to describe the three stages which should occur in each investigation – the exploratory, initiated by teacher or children, in which the teacher interferes as little as possible, asking questions only to assist progress; a second stage where the teacher focuses attention on salient points which are considered important; and finally a follow-up to consolidate what has been learned. The reflections which follow are based upon a wealth of experience of mathematics learning at this level both in this country and overseas.

There are certain points which must be emphasized about learning by this method, for there has been much misunderstanding about so-called 'discovery learning':

1 Practical work is time-consuming, and teachers must always have a clear mathematical aim in mind for each activity undertaken by the children. As wide a variety of mathematical ideas as possible should be abstracted from any one investigation

2 Pupils do not normally learn from one discovery alone. It is therefore essential to plan progressive experiences in important topics. These topics should be re-introduced at intervals, each time at a more demanding level than before

3 All pupils do not require the same kind of experience in order to learn a particular mathematical concept. Some need extensive and varied experiences with real materials, while others soon put materials aside and investigate in a more abstract way. Investigation does not necessarily require real materials. Sometimes a solution to a problem is a pattern or relation, and paper and pencil are required

4 The work should be planned to encourage pupils to investigate and experiment for themselves. Their suggestions should be followed up, and they should not feel restricted by their teacher's programme

5 It follows that teachers need to keep careful records of children's achievement and progress. (ibid., p. 9)

9 Primary mathematics: Nuffield and after

The Nuffield Primary Mathematics Project was set up in September 1966 with Professor G. Matthews as organiser. Professor Matthews, formerly Head of Mathematics at St Dunstan's College, had been one of the group of individuals who in the late fifties and early sixties had been independently active in trying out new content and approaches to secondary mathematics in their own schools (Rollett 1963). In many cases these activities had not been part of any formal project, though Professor Matthews and his colleagues did ultimately produce a series of topic- and text-books, with an O level syllabus and examination administered by the Oxford and Cambridge Board (later amalgamated with M.E.I. O level).

The original intention of the project was to examine the teaching of mathematics in the age range 8–13. (At that time a multitude of secondary projects was springing up, covering roughly the 11–18 age range.) It was soon decided that there would be advantages in considering the development of mathematical ideas from the beginning of a child's school life, and so the age range was extended to 5–13. The project followed squarely in the tradition, already established, of learning by doing and adopted as its motto a 'Chinese proverb' (widely believed to have been specially concocted at project headquarters!):

> I hear, and I forget
> I see, and I remember
> I do, and I understand

The first publication of the project was a description of the aims and teaching methods envisaged, entitled *I do and I understand* (and dedicated to Jean Piaget). The following quotation is from the introduction:

The aim of the Nuffield Mathematics Project is to devise a contemporary approach for children from 5 to 13. The guides do not comprise an entirely new syllabus. The stress is on how to learn, not on what to teach. Running through all the work is the central notion that the children must be set free to make their own discoveries and think for themselves and so achieve understanding, instead of learning off mysterious drills. In this way the whole attitude to the subject can be changed and 'Ugh, no, I didn't like maths' will be heard no more. To achieve understanding young children cannot go straight to abstractions – they need to handle things (apparatus is too grand a word for at least some of the equipment concerned – conkers, beads, scales, globes, and so on).

But setting the children free does not mean starting a riot with a roomful of junk for ammunition. The change-over to the new approach brings its own problems. The guide *I do and I understand* (which is of a different character from the others) faces these problems and attempts to show how they can be overcome. The other books fall into three categories: *Teachers' Guides, Weaving Guides*, and *Check-up Guides*. The *Teachers' Guides* cover three main topics: Computation and Structure; Shape and Size; Graphs Leading to Algebra. In the course of these guides the development of mathematics is seen as a spiral. The same concept is met over and over again and illustrated in a different way at every stage. The books do not cover years or indeed any specific time, they simply develop themes and therefore show the teacher how to allow one child to progress at a different pace to another. They contain direct teaching suggestions, examples of apparently un-mathematical subjects and situations which can be used to develop a mathematical sense, examples of children's work and suggestions for class discussions and out of school activities. The *Weaving Guides* are single-concept books which give detailed instructions or information about a particular subject. (Nuffield Foundation 1967)

Initially all the publications were addressed to teachers – only in the later stages was material for pupils produced. This was aimed at pupils in the top forms of middle schools or the first two years of secondary schools and was in the form of twenty 'modules', sets of work cards for pupils, with an accompanying teachers' booklet, each capable of standing alone and available to supplement other books and material. The teachers' guides were intended as suggestions out of which ideas might be developed, rather than a prescriptive course. 'They were written against the background of teachers' centres where ideas put forward in the books could be discussed, elaborated and modified' (Nuffield Foundation 1967, Introduction). The teachers' centres were an important feature. The local authorities which acted as pilot areas also agreed to provide these centres as part of their commitment to the project. This gave a considerable impetus to the development of teachers' centres in general and was a major step forward in in-service provision for teachers.

Fourteen pilot areas were chosen, spread over the country; there was heavy pressure from other local education authorities to be involved, so that seventy-eight areas were designated as 'second phase' and forty-four more as 'continuation' areas. The guides appeared first in draft form and were modified after discussion with teachers who tried out suggestions in the classroom. In all, eleven *Teachers' Guides* and six *Weaving Guides*, were published. A series of *Check-Up Guides*, prepared in association with the project by members of Piaget's team, under the supervision of Dr L. Pauli of Geneva, has also been produced. These are intended to help teachers to confirm progress in concept attainment made by individual children. Other publications (apart from the twenty modules already mentioned) were *The Story so Far*, later superseded by *Guide to the Guides*, *Maths with Everything* (illustrating how opportunities of mathematical experiences for 5–7 year olds might be utilised), and three sets of *Problem Cards* (Green, Purple, and Red, in order of increasing difficulty) for use with young secondary children.

Three films were also produced illustrating the work of the project: *I do and I understand* – at junior level; *Into Secondary School* – at lower secondary level; and *Maths with Everything* – about infants. Five television programmes, *Children and Mathematics*, made with the cooperation of the B.B.C., were also used to publicise the work, and a magazine, the *Nuffield Mathematics Teaching Project Bulletin* was circulated via local education authorities. This contained teacher comment, suggestions for classroom approaches, and news, and has continued an independent existence as *Mathematics Teacher's Forum* even after the conclusion of the project.

A major problem was that of in-service training for the project teachers. The Department of Education and Science, members of the project team in their trial areas, and other independent bodies were all involved in the provision of courses. Annual Conferences were held in Leicester, involving the writing team and area organisers, so that feedback from the trial areas could be discussed. Team membership gradually changed over the years; at its height in 1966–7 there were seven writers plus the organiser. When it finally drew to a close in 1971, the project could fairly claim to have made a major contribution to developing and extending the changes in primary mathematics education which had preceded it. Large numbers of primary teachers had attended courses on mathematics teaching and in many schools the suggestions of the guides were being taken up. A network of teachers' centres was now springing up and new series of pupil texts were appearing developing the Nuffield ideas.

Fifteen years after the publication of the Mathematical Association report, during which time primary mathematics had changed almost beyond recognition, the association published a further report (Mathematical Association 1970). Comparison of the two shows how ideas which earlier seemed revolutionary were now accepted as commonplace. Yet many problems remained:

Re-examining the 1955 report we find no major principle that we wish to retract; there are, however, a few statements in which we would wish to modify the emphasis, the scope or the setting. Principles that were then confidently stated and perhaps too easily accepted are seen now to involve more difficulties than were foreseen. (Mathematical Association 1970, p. 4)

(The growth of children's understanding and the problems of trying to assess it are mentioned as a case in point.)

Though for the most part, the changes have been related to method rather than the introduction of 'modern' content, (work on shape, graphs, and simple algebraic ideas, though 'new' to the primary school, is not modern in this sense), some additional topics have been introduced – in particular those of 'sets', spatial ideas such as vectors and tessellations, and number systems (integers, algebraic structure, etc.). These have sometimes led to difficulties; they were totally unfamiliar to many teachers, and they have occasionally been pressed too far by teachers who were unsure of their significance, but felt that 'modern mathematics' demanded this of them. A reaction against an unduly deferential attitude to 'what is important in mathematics' may be found in an excellent report of the Association of Teachers of Mathematics:

It can be argued that the teaching of mathematics has concentrated exclusively on the attempt to transmit known mathematics, and that it has done this in an incredibly narrow way. Our view of mathematics makes us see the job of the teacher differently. We are concerned with the creative side of the child's learning and with minimising the teachers' interference with this. Every time a teacher insists on his way of doing a piece of mathematics, rejecting any responses which do not seem to fit, he nibbles away at his pupils' ability to act mathematically . . .

Primary school mathematics, though, is not only about numbers. We approve the teaching of some plane and solid geometry, the use of coordinates, graphs, simple numerical algebra, and some topics usually called 'modern': sets, simple examples of algebraic structures, elementary topology, and so on. But what do these words mean? . . . 'sets' may stand for the manipulation of Venn diagrams to solve pseudo-problems; 'graphs' may mean a formalised routine for illustrating relationships; if these are the meanings the words carry in practice, we want nothing of them. (Association of Teachers of Mathematics 1967, pp. 4, 6)

This book, which aims 'not to provide a programme for primary school mathematics but to stimulate experiment', has many suggestions for interesting topics and many illuminating insights on the purposes of teaching. (Though some of its suggestions might prove rather ambitious with young children, the book is a mine of ideas.)

Underlying this debate lies a basic problem which reappears at all levels in mathematics teaching, the possible incompatibility between the interests and background knowledge of the learner, and the logical sequence which, it is argued, is inherent in the nature of mathematics itself. If children engage unrestrictedly in 'child-centred investigation' of unrelated topics, they may acquire an uncoordinated hotch-potch of ideas which by its lack of any coherent structure may ultimately impede their progress.

Systematisation without a basis of ideas springing from experience has been a feature of much mathematical education in the past. Experience which children never build into an inter-related body of ideas would be an equally unfortunate development. (Williams and Shuard 1970, p. 1)

Concern has been expressed by some observers at the lack of sequence in primary school mathematics. What is at issue here is not the tidy logical exposition of a body of developed theory, appropriate to a mature sophisticated review of the subject at an advanced level, but the psychological links – and orderings – of the topics which the child meets.

One can find out, adequately enough, what domestic life was like a century ago without first investigating life a hundred year's earlier, even if the development through the century is then less clear and significant, but one cannot master the process of division without first being familiar with multiplication. (Gardner and others 1973, p. 120)

In response to this problem, schemes of work have been suggested – a recent one was produced by West Sussex teachers (West Sussex County Council 1975). Indeed this problem was recognised in the works of the Nuffield project – the second half of *Guides to the Guides* contains selections from a number of outlines and schemes of work produced in teachers' centres – though it is emphasised (particularly in the 'Anti-introduction' by A. J. McIntosh, himself a mathematics adviser) that the suggestions are not intended to be prescriptive:

The Nuffield team have written this guide *specifically to help people make up schemes of work for themselves*, to suit their own situations. Their stated intention is that you should *not* take over any of the schemes of work, in this book: 'bespoke tailoring rather than off the peg' is their way of putting it. (Nuffield Foundation 1973, p. 55) [my italics]

Most teachers would agree that some sort of scheme of work, outline, or framework is necessary if we are to plan our approach to mathematics teaching. Disagreement starts, however, when we try to define what we mean by a scheme or an outline, and to decide how detailed or deliberately vague it should be. Opinions range from 'No scheme at all – let the mathematics emerge naturally', to 'I want to know exactly what these children did last term and just what they are going to do, week by week, during this term.' Between these two extremes – the 'Let-it-emerge League' and the 'Strait-jacket Squad' – we would expect to find the vast majority of teachers who, whilst requiring some direction and over-all plan for their teaching, do not wish to be tied to a line-by-line syllabus. The problem would seem to be how to produce a scheme which is structured enough to be of help without its being so rigid that it stifles the ingenuity, imagination, or 'flair' of the individual teacher. (ibid., p. 53)

Some aspects of the Nuffield approach to computation have been criticised for their 'over systematic' treatment based on the use of sets, mappings, and ordered pairs (figure 9.1); there is a danger that in attempting to formalise and make rigorous the treatment, the texts will encourage an over-sensitivity to the structures of formal mathematics at the expense of children's own mathematical activity.

Mathematics in primary schools faces other problems, too, an important one being the fact that nearly all primary teachers are 'non-specialists'. In some cases, unfortunately, teachers who as pupils were uniformly unsuccessful at mathematics, and who have developed reactions of fear and distaste towards the subject, find that they have, willy-nilly, the responsibility of guiding the mathematical development of their pupils. Publicity has been given from time to time to the low level of formal mathematical qualification of many college of education entrants. Colleges often regard part of their work in 'professional' or 'curriculum' courses in mathematics teaching-method as an attempt to influence attitudes for the better, and dispel some of the fear felt by students. In some cases qualifying courses in basic mathematical proficiency have been instituted in a bid to raise standards during the course. Although these have necessarily been pitched at a very low level, they have sometimes induced even more fear and resentment among some of their clientele. Demanding better

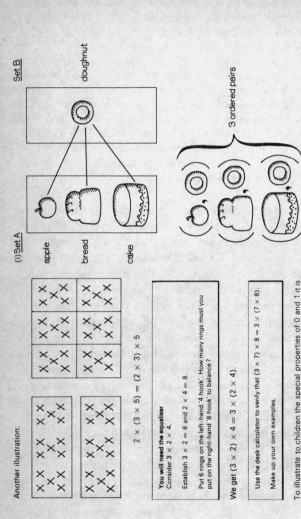

(i) Set A   Set B

apple

bread

cake

doughnut

3 ordered pairs

Another illustration:

$2 \times (3 \times 5) = (2 \times 3) \times 5$

You will need the equaliser
Consider $3 \times 2 \times 4$.

Establish $3 \times 2 = 6$ and $2 \times 4 = 8$.

Put 6 rings on the left-hand '4 hook'. How many rings must you put on the right-hand '8 hook' to balance?

We get $(3 \times 2) \times 4 = 3 \times (2 \times 4)$.

Use the desk calculator to verify that $(3 \times 7) \times 8 = 3 \times (7 \times 8)$.

Make up your own examples.

To illustrate to children the special properties of 0 and 1 it is advisable to return to products as on p. 25, e.g.

The cardinal number associated with Set A is 3 and the cardinal number associated with Set B is 1. The cardinal number of the product set is $3 \times 1 = 3$.

Figure 9.1   Extract from the Nuffield guide *Computation and Structure 3* (Chambers and Murray, 1968, p. 29)

initial qualifications from the smaller number of students who will now be entering teacher training over the next few years may be the way forward. It is still the case, however, that a teacher, once 'certificated' as a secondary specialist in music and history, with minimal preparation in mathematics, may yet find himself (or herself), through force of circumstances, teaching general subjects in a primary school.

Though some primary schools are now encouraging individual members of staff to take an overall responsibility for particular subject areas such as mathematics, most are still 'generalists'. One happy feature of this lack of specialism has been the willingness of the teachers to learn and to adopt new approaches. Their voluntary attendance at in-service courses has been noteworthy, and where their secondary colleagues have seemed at times unable or unwilling to recognise their need for updating, primary teachers have come forward with enthusiasm and refreshing openness to new ideas. In view of the immense changes taking place in other primary school subject areas, as well as in overall patterns of school organisation, this is the more remarkable.

Clearly the demands made on the teacher are almost unlimited. She must not only have considerable understanding of the foundations of mathematical thinking and learning and the inter-relations of its different aspects, but also she must be able to cope similarly with the vital early stages of language work, movement, creative activities, music, and all the rest. She is expected to recognise growing points in all these areas and to lead perhaps forty or more youngsters through this maze in steady growth.

So we are brought up against two apparently conflicting requirements for the teacher – the depth of specialised knowledge on the one hand and the width of experience and broad array of subject matter to be covered on the other – and each of these requirements today goes far beyond the bounds set in most schools even as late as the middle of this century. (Gardner and others 1973, p. 6)

Another problem faced more acutely in primary than in secondary schools arises from the 'integrated' approach to learning. Gardner and others (1973) give an excellent discussion of this issue, referring to the 'doubts and difficulties seen by many experienced and compe-

tent teachers'. Asking 'exactly what kinds of school organisation and
activity do people have in mind when they use phrases such as integrated day, topic approach . . .?', the authors go on to enquire what
the place of mathematics is in each of these, and whether it is possible
to 'gain an adequate and eventually a systematic knowledge of
mathematics from such approaches' (ibid., p. 116). They take as an
example the theme of 'communication', from which might arise such
topics as: travel in the Middle Ages; the story of the telephone;
writing through the ages; secret codes.

We notice at once that mathematics does not arise naturally from the list, as
work slanted towards other traditional subjects appears to do. The mention
of codes has been made deliberately because it does suggest, but only to a
mathematically informed reader, such topics as permutations, frequency
distributions, encoding matrices, and the like, all of which can only be effectively developed at post-primary levels. They certainly go beyond the basic
and essential stages which the primary child must somehow cover.

. . . the integrated day if not carefully planned and sensitively supervised,
may leave large gaps in knowledge, may leave essential concepts unformed,
and may fail to achieve any sort of progression. Where facility depends on
practice, it may well hinder progress through failure to provide it. On the
other hand, the integrated day could provide just that free atmosphere and
stimulating situation in which children best thrive.

One does not doubt that many topics are rich enough in mathematical
ideas to meet the needs of the primary school child. Many topics could indeed generate mathematics if only the teachers were sufficiently aware of
the possibilities. (ibid., pp. 117, 119, 120)

The authors conclude that even where integrated work is functioning
smoothly and successfully, time should be set aside for specific work
in mathematics, with the objectives of practising skills (need for
which may have arisen in a project), filling gaps in knowledge, re-
consideration of the work already done so that children can grasp its
development, and extending work mathematically so that knowledge
becomes systematic. 'Until it becomes systematic, it is not
mathematics.'

The changes which have taken place in mathematics in primary school are perhaps even more sweeping and widespread than those in the secondary schools; the present pattern has been investigated by a Schools Council Project (Primary School Mathematics – Evaluation studies) headed by Professor J. Wrigley and based at the University of Reading.

The project aimed to gain a general picture of the ways in which primary school mathematics is currently taught, and investigated children's competence in mathematics as well as teachers' attitudes and methods. An account on some of its findings was given by Ward (1974); forty mathematics questions on a range of topics were given to 867 ten-year old children in twenty-five classes. Their performance was compared with teachers' ratings of the importance of each question – computational skills, practical applications (e.g. shopping), notation, and place value were rated highly, the last being a topic where the children were relatively weak. Although only sixteen schools from three local education authorities were involved and the sample was not random, this pilot survey gives an interesting picture. Other information on organisation and methods was also collected. Two of the schools were streamed and three used setting by ability for mathematics; most had a written scheme of work and used a mixture of class group and individual work; *all* practised tables (take heart, industrialists!) and seemed well equipped with a variety of apparatus. The teachers felt mathematics was now more enjoyable for children. They emphasised the role of 'understanding', were watchful to prevent decline in computational skills, and expressed concern about possible lack of progression and continuity in the work. 'The child can meander aimlessly through a lot of work and draw few conclusions, learning little.'

A larger survey involving a stratified sample of forty schools from twelve local education authorities ('old style') confirmed the general picture. This time 7 per cent of classes did not practise tables. About a quarter of the schools were streamed or setted for mathematics and half used grouping by ability within mixed ability classes. Information on textbooks in use, designated specialist teachers (only 20 per cent, depending greatly on L.E.A. policy), and specialist teaching rooms (none! – one school teaches two classes in a cellar) was compiled. There was considerable variation in the amount of time spent on mathematics each day (mostly about one hour, but ranging from 30–90 minutes).

Because of all the changes that have taken place in the teaching of the subject in the last few years, it was felt that many teachers might find it helpful to have a general picture of what is going on in primary mathematics. In this survey we try to give such an overview. No attempt has been made to compare different methods or to give a Best Buy; even if this were desirable it would not be possible because there must be very few schools who stick to one approach or one series of textbooks, to the exclusion of all others. We have, however, described the great variety in method and outlook that was found. (Schools Council 1974)

The School Mathematics Project has also turned its attention to the younger age groups by forming a working party to produce materials for middle schools. This was launched at a four-day working conference in spring 1972 attended by over a hundred participants. The deliberations of the twelve groups on a variety of topics resulted ultimately in a seventy-six-page report (School Mathematics Project 1972). The aim is to produce material for the 7–13 age range, in three differently-paced versions, steering 'a middle course between a relatively inflexible textbook series on the one hand, and unconnected and unstructured materials on the other (by) a judicious mixture of topic books, work cards, work sheets and supplementary material' (Rogerson 1973).

A team of twenty-five writers, mostly teachers, was assembled in September 1973 and by spring 1974 had produced material for the 7–9 age group. The teachers' handbooks and supplementary books include course outlines, concept maps, answers, and help on organisation and materials, and individual pupil record cards, linked to the pupil materials, are provided. Many of the ideas had already been used by the writers in their schools; the draft material, attractively designed in a variety of styles, is being tested during the two school years 1974–6 in twenty-nine schools. Information collected in the formative evaluation will be used to produce a final version for publication.

Aims and objectives

1 We wish to write a complete mathematics course. The data will indicate whether the materials are too much/too little and what areas
need supplementation

2 The course should have a logical progression of mathematical activities
and concepts and the feedback should enable us to make adjustments
where necessary

3 Each child should be respected as an individual, and material should be
designed to give him satisfaction and enjoyment in achieving his
potential

4 The course should include group work and encourage occasional class
teaching as well as individual work so that a child will cooperate with
others and feel a part of a larger social group

5 The mathematical activities should relate to the child's own life and environment where relevant

6 Children should be encouraged to use their intuition in creative,
experimental, and discovery work

7 Children should use apparatus frequently and engage in concrete activities where relevant

A large number of teachers and others are involved in helping to make S.M.P.
7–13 an appropriate and useful contribution to mathematics in the middle
years. Systematic in-service help will be provided including courses,
explanatory guides for teachers' centres and colleges of education, and video-
tapes to explain S.M.P. 7–13. Another important area will be the preparation of
tapes (and possibly slides) for remedial use. (Rogerson 1975, p. 6)

10 The Mathematics for the Majority Project

Planning

Mathematics for the Majority Project was a development which broke new ground in several respects: it was a deliberate attempt to cater for the 'non-academic' pupil (in contrast to the large number of O level/C.S.E.-oriented schemes which formed the first wave of curriculum development in secondary mathematics); it was one of the first projects in which 'evaluation' was built-in from the start; and it was initiated and funded by the Schools Council. The project had its origins in a feasibility study commissioned by Schools Council in 1965, carried out by Floyd. The gradual spread of comprehensive re-organisation was forcing recognition that thought must be given to the curriculum of the average pupil, and this study, examining the mathematical needs of the less able and how these might be met, was published by the Schools Council (1967) — *Mathematics for the Majority*. Following this report, the Schools Council Secondary School Mathematics Project was set up, with Floyd as director, based at the University of Exeter. Its aims were described as follows:

The project aims to meet the mathematical needs of pupils of average and below average ability between the ages of thirteen and sixteen. It is concerned with the production, trial, and dissemination of teaching materials, presented mainly through a series of guides for teachers. The aims of the project team are: (a) to provide pupils with experiences of mathematical situations, to encourage powers of judgment and the exercise of im-

2. We raise the question of the measurements a builder would have to make in order to draw a plan of the field to use in his office. Some pupils may suggest that the lengths of the four sides would be sufficient. We may, therefore, decide to measure the sides and return to the classroom to draw our own plans.

3. Sooner or later it should emerge that these measurements alone are not enough to allow us to draw an accurate plan since no two pupils will produce identical plans. In particular the question of the *angle* which two lines must make cannot be answered with the data we have. At this point the teacher produces a framework of four strips of wood which he has previously prepared: this framework can be distorted but in one position it will be a scale representation of the plot.

4. When pupils see that they can distort the framework to make many different shapes, they may suggest that:

   (a)   another strip 'across the middle' (Fig. 28) will make it *rigid* (a new word?);

Fig. 28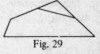

   (b)   four measurements are not sufficient;
   (c)   a strut or diagonal strip corresponds to a fifth measurement which has to be made outdoors;

and even perhaps

   (d)   both diagonals should have been measured;
   (e)   a diagonal is not the only other measurement that could make the framework rigid (Fig. 29).

Fig. 29

5. Somebody can be sent to measure the appropriate diagonal, and a fifth strip, cut to scale, can now be added to the framework which will then be 'the same shape' as the plot.

6. We may need to discuss such questions as: Ought we to have measured the other diagonal as well? or instead? and: If we measure one diagonal, shall we now be able to read from our plan the actual length of the other diagonal?

This can be the beginning of an investigation in which every pupil makes rigid frameworks from strips of stiff card, deciding how many diagonal or other struts are necessary, and where they should be placed. Typical questions to ask might be:

   Are *all* possible diagonal struts *necessary?*
   What patterns appear when we list the number of sides/diagonals/ triangles formed?
   Can we make generalisations (in *words,* not symbols)?
   What conclusions do we draw about the number of measurements needed to make plans of fields with 5, 6, 7 . . . sides?

Figure 10.1   Extract from *Mathematics from Outdoors* (Chatto and Windus, 1972, p. 27)

agination; (b) to give pupils some understanding of the mathematical concepts which underlie the numeracy required for everyday affairs; (c) to remove barriers isolating mathematics from other areas of the curriculum and other interests of the pupils; (d) to enable pupils to appreciate in some measure the order and pattern of their environment. (Floyd 1969, p. 4)

Initially funded for the three years 1967–70, the project was later extended until 1972 as it soon became apparent that the programme as originally conceived could not be completed in the time available. Its advent was welcomed by many mathematics teachers who, aware of new developments in the secondary field, felt these inappropriate for their less able pupils. They hoped now that material would be provided which would help them with emerging problems in this area, particularly as the oft-heralded and much postponed raising-of-the-school-leaving-age was in one of its 'go' phases. (It became a reality in 1973.)

A number of initial decisions were made which were to have profound consequences for the outcome of the project. Materials were to be produced for teachers (as a series of 'guides') but not for pupils; in some cases the guides were to be written by practising teachers with firsthand experience of the problems involved. The emphasis was to be practical rather than academic, with relevance and pupil motivation as important criteria. Mathematics classrooms, with a variety of equipment, and teaching in double periods to allow practical activity were advocated. Subject boundaries were blurred, or ignored; the work was to be pupil- rather than subject-centred (see figure 10.1). In this, as in its advocacy of group work and pupil assignments (work cards), the project was following the trend already established by the reform in primary schools. A new role for the teacher was implied:

Information and instruction are given by staff who see themselves (and apparently are accepted by the children) as members of the team, guiding the work rather than pontificating. This change of role in the teacher has established a change in relationship between teacher and pupil. There is increased communication . . . it has been much easier for a child to admit lack of comprehension. (Floyd 1969, p. 5)

It was envisaged that teachers would use the guides to prepare their own materials, adapting the suggestions specifically to the needs of their own pupils; in a favourite project phrase, 'the teacher knows best'.

It is no part of the Director's policy to provide prescribed courses of action, let alone a syllabus or a pre-packed kit of mathematical tools. (ibid., p. 5)

It was suggested that in-service training be based on teacher discussions and interchange visits rather than the exhortations of 'experts'.

The large project budget of £83,000 defined it as a major development. Almost immediately it faced considerable problems of logistics. Over 100 local education authorities were involved, represented by 26 'pre-pilot' schools, 87 'pilot' schools, and 280 'associated' schools. Communication was maintained by means of *News Sheet* (appearing 'as occasion demanded' – roughly at six-monthly intervals), by visits of the project team to the pilot and pre-pilot schools, and by using local education authority advisers, together with college and university tutors, as 'liaison officers' to disseminate information to the associated schools. The intention was to subject the first drafts of material to criticism in the pre-pilot schools, and then to prepare trial versions for use in the pilot schools. The associated schools which had expressed interest and wanted information about the project could be supplied only with single copies of the guides; they were too numerous to be used as pilot schools, and had to be left, in the main, to their own devices. The liaison officers were kept up to date on developments by a series of national conferences and they in turn arranged local conferences and meetings for teachers from associated areas; often these also involved pilot schools, and sometimes project staff (one of whom had special responsibility for links with associated areas).

Feedback

As the production of guides began arrangements were made for collecting feedback information from pilot schools. Questionnaires and invited comments on the teachers' guides were intended to give immediate teacher-reaction to their content, but additional information was needed on their effectiveness in the classroom; sets of pre- and post-test tasks were prepared which would indicate the extent of pupils' mastery of the material covered. The help of college of education tutors was enlisted in the data collection — each tutor was assigned to work with one or two classes; pupils completed the test tasks to identify progress during the year and were also interviewed to gauge their understanding of the mathematical ideas underlying the practical activities suggested by the project. Interviews with teachers were also employed to highlight difficulties related to the method and materials. Some sixty-two classes in thirty-six pilot schools were chosen for these more detailed 'case studies'. Initial data were collected on third- and fourth-year pupils in these schools (on I.Q., attitude to mathematics, personality characteristics, ability as measured by an N.F.E.R. 'basic mathematics' test, and reading ability). Later measurements showed how these changed with time. (Attainment level appeared to increase over the two-year period, attitude showed more fluctuation with considerable variation from school to school (Project *News Sheet*, 5 (1971)).

One very striking finding from the initial data concerned the allocation of specialist mathematics teachers to pupil groups of low ability, which, taken together with the decisions of principle made by the project, has far-reaching implications.

Teachers were categorised as:

1 Mathematics specialists
2 Other specialists teaching full-time in mathematics department
3 Other teachers (those teaching part-time in mathematics department, head teachers, remedial, etc.)

Pupil groupings were:

1 Academic
2 Average
3 Below average

The results were:

Mathematics teacher categories by pupil sets (whole sample)

Fourth year		Teacher categories			
		1	2	3	Total
		%	%	%	%
	1	84	8	8	100
Pupil groupings	2	69	10	21	100
	3	45	6	49	100

Third year		Teacher categories			
		1	2	3	Total
		%	%	%	%
	1	85	9	6	100
Pupil groupings	2	67	10	23	100
	3	41	6	53	100

There were marked regional variations, but Kaner summed up the situation:

Over half the students in the third category are taught mathematics by non-specialist teachers, and it is clear that improvements in the standard of the mathematical education of these students are largely dependent on the training and assistance given to such non-specialist teachers. This fact is vitally relevant to the Mathematics for the Majority Project. (Schools Council 1970, pp. 26–7; table slightly adapted)

Outcomes

During 1967 the project team was appointed, and arrangements made for participation of schools in pre-pilot, pilot, and associated areas. During the following year conferences were held for teachers from pilot areas and for liaison officers, and work began on the writing of the teachers' guides. Getting the ideas down on paper and casting the material into a suitable form took longer than expected, however, and there were delays in getting initial feedback, in production and distribution so that the first draft guide was not available until May 1969.

A series of discussion questions (p. 83) was suggested in the hope that teachers would give preliminary thought to such issues as working in groups and the use of assignments, the impact of such innovations on the role of the teacher, and their relationship to external examinations.

Logistic problems continued to affect the project, despite the relief afforded by a two-year extension of its working; once production of the guides had fallen behind schedule it required a tremendous effort to catch up. Communication problems were considerable, too, with such a large number of schools involved. Feedback, perhaps inevitably, was slow in coming – teachers receiving the guides needed time, and often help, to produce materials from them. There were a few instances where teachers mistook the evaluation tasks for project work cards and used these in their teaching, no doubt to the consternation of the evaluators! (*News Sheet*, **3** (1970) p. 6). Only after teachers had produced and tried their materials could they return their comments on the guides to the headquarters team in Exeter, writing the next batch of guides, and wondering impatiently what were teachers' reactions to the earlier ones.

The project staff gradually became depleted as terms of secondment came to an end; for the last twelve months of the project's life the central team consisted of only three members. By this time Schools Council had already set up the Mathematics for the Majority Continuation Project, also based in Exeter. This was directed by P. Kaner who had been the Evaluator of the original project, and aimed to carry forward the work by producing pupil materials. For a period the two projects worked alongside each other.

Despite all these difficulties and changes, a series of fifteen teachers' guides was eventually produced; several of the early trial versions were extensively revised on the basic of teachers' comments and additional writers were involved in the work.

The guides, all published by Chatto and Windus, who had produced the draft trial version, are listed on p. 82. Some of these presented fairly traditional material, often through a practical approach – *Mathematics from Outdoors* (surveying and navigation); *Some Simple Functions* (linear, square law, periodic, exponential); *Algebra of a Sort* (pattern generalisation and formulae); *Number Ap-*

preciation; Machines, Mechanisms, and Mathematics (an unorthodox approach to geometry – see p. 80). Of the remaining books, some were concerned with teaching method and the aims of the project, and others introduced 'new' material in an original and stimulating way – probability, calculators, simple geometrical and topological puzzles (*Geometry for Enjoyment*), and investigation of pattern. Mechanics, graphs, computers, and flow-charting were included in *Space Travel*, while *Crossing Subject Boundaries* examined the inter-relationships of mathematics with geography, science, handicrafts, art, music, sport ... even history (radioactive dating, family trees, population statistics, weapons, architecture) – a useful source of suggestions for broadening and integrating the work.

Overview

What then was the outcome of the Mathematics for the Majority Project and what lessons are to be learned from the experience gained? There can be no doubt of the project's timeliness – indeed, the very fact that for so many schools it promised to cater for a deeply-felt need was to some extent its undoing. Looking back one may wonder whether a more ruthless approach, cutting down the number of pilot schools, and, at least initially, not acknowledging associated areas, might have reduced the logistic problems. Naturally, the team were concerned to gather ideas, useful current practice, and teachers' opinions, and the idea of spreading the net as wide as possible is an appealing one – but it now appears that many of the teachers concerned, maybe mistakenly, saw the project only as a provider, and not as a collector/disseminator of ideas.

Perhaps the most significant factor in this situation is one, already referred to, which emerged from Kaner's evaluation data – the large proportion of non-specialist teachers who were involved in teaching the pupils whom the project was aiming to help. Apart from the limitations of time and circumstances which hamper even the keen and mathematically-qualified teacher wishing to prepare materials for his pupils' individual use, many of the teachers involved must have found that their lack of mathematical background placed them in a peculiarly difficult position. For example, the guide *Machines, Mechanisms, and Mathematics* showed, in an original and stimulating way, how much of the content of elementary geometry could be examined and illustrated through a study of quite simple everyday 'machines' – gear wheels, park playground furniture, trellis fencing, and the like. One can sympathise however, with the geographer, P.E. teacher, or deputy head who, not entirely *au fait* himself with the geometry involved, and maybe reluctantly obliged to teach a few periods of mathematics each week, is confronted with the guide and realises that he has to:

1 Understand the mathematics background
2 Perceive its realisation in the situations presented
3 Write assignments, which his pupils can understand, based on this material

Clearly the task is formidable. Here was the prime reason for the institution of the Mathematics for the Majority Continuation Project.

It might be thought that, once the guides were produced, the task of the project was complete. But as Floyd wrote:

We conclude with a reminder of the stark fact that the project officially will cease to exist on 30 April 1972. After that date there will be no project staff and no project headquarters. Thereafter the project must live as through the medium of the published work, and more significantly through the actions of teachers, whether working as individuals or in local consortia. (*News Sheet*, 5 (1971))

Thus, just at the point when the project team had material to display and experience of its use to pass on, the team was in process of being disbanded! (This problem has not arisen in the case of S.M.P. Though the intention was that it would complete the work by 1968, its continued activities, now self-supporting, have meant that the accumulated expertise, with an organisation to support dissemination, has remained in being.)

In response to requests for help and suggestions from pilot and associated schools, Floyd and his colleagues produced a 'Teachers' ideas and resources kit' – of discussion sheets and suggested work cards based on three of the guides. These were intended for teachers' centres and working groups of teachers, and proved useful in the conferences for associated area schools and the wider dissemination conferences which were mounted independently of the project by local education authorities and institutes of education. A set of colour transparencies with taped commentary, showing pupils at work on trial materials, was also made available on loan.

More recent Schools Council projects have reflected the realisation that unless the materials are made widely available to teachers, much of the effect of a development will be nullified. Greater attention is now paid to dissemination, and latterly extension grants have been given by Schools Council to projects in order to support the provision of conferences and other means of making the material and advice on its use available to teachers.

The Mathematics for the Majority Project quite consciously built upon the earlier work of the Nuffield 5–13 Project, in which some of the teachers concerned had previously been involved. The ideas of teachers' working groups and activities in teachers' centres were incorporated in the approach. Perhaps more fundamental (at least for teachers who had not had such prior experience) was the new way of working which was suggested (see the quotation, p. 73, about the 'change of role in the teacher'). 'The reactions of those teachers who have broken the back of the transition from traditional methods naturally differ from the reactions of those without experience of "the discovery approach" as applied to mathematics' (Floyd 1969, p. 4).

This new pattern of classroom activity, implicit in the types of pupil activities, involving measurement, use of equipment, individual assignments, group 'probability' experiments, etc., was only just beginning to find its way into secondary schools at that time. Today the situation has changed somewhat and new ideas are more familiar – the use in primary schools of less formal groupings is better known, and many secondary schools, perhaps after undergoing re-organisation, are seeking new approaches and teaching methods. The Mathematics for the Majority Project no doubt played its part in encouraging teachers to examine alternative approaches –

but where these were unfamiliar, they acted as yet another impediment to the acceptance of its materials. (These and other issues were raised in a document 'Suggestions for discussion topics', part of *News Sheet* 2, extracts from which are given on p. 83).

But one is bound to add that a project which tried to break completely new ground in the teaching of the non-academic child would inevitably encounter severe difficulties. This was a daunting task. The O level reforms, arising mainly in grammar and public schools, were geared to children who were generally capable and well-motivated; the problems of teaching mathematics to such children had been fairly thoroughly explored during the first half of the century (for example, in the reports of the Mathematical Association already referred to).

By contrast, the non-academic pupils had generally left school early, probably having encountered only arithmetic and a little geometry. The problems of what mathematics might be appropriate for the majority and how it should be taught, had received little attention until now, apart from *Mathematics in Secondary Modern Schools* (Mathematical Association 1959). A development project in this area was bound to run into difficulties of a much deeper nature than its predecessors, formidable difficulties which even yet are largely unresolved.

Appendix

1 Publications of the Mathematics for the Majority Project
   (Chatto and Windus, 1970–4)
   Mathematical Experience
   Machines, Mechanisms, and Mathematics
   Assignment Systems
   Number Appreciation
   Mathematical Pattern
   Mathematics from Outdoors
   Luck and Judgment
   Space Travel and Mathematics (vol. I)
   Space Travel and Mathematics (vol. II)

Algebra of a Sort
From Counting to Calculating
Crossing Subject Boundaries
Some Simple Functions
Some Routes through the Guides
Geometry for Enjoyment

2 Some Suggestions for Discussion Topics
(Extracts from The Schools Council Secondary School
Mathematics Project *News Sheet*, **2** (1969))

AIMS AND OBJECTIVES

(a) Of the many reasons put forward for including mathematics in
a pupil's curriculum, what are seen as priorities?

(b) Do these priorities vary (in existence, intensity, or relative importance) according to age? To general ability? To mathematical
ability?

ORGANISATION

(a) Can a normal-sized class be taught as a whole class all the
time? Or some of the time? If not taught as a full class, how it is to be
organised? Do decisions taken about this depend on the school's
organisation, e.g. on there being or not being systems of streaming or
setting?

(b) Can any form of class organisation ensure a close homogeneity
of ability and response? To what extent might this depend on the size
of the class?

(c) To what extent is discussion (on proper topics!) among the
pupils, as well as between teacher and pupil, to be encouraged? All
the time? Much of the time? Very occasionally? Never?

(d) Is movement about the classroom (and also in and out of it)
allowed? Freely? Strictly limited?

SYLLABUS

(a) Is a written mathematics syllabus necessary? Desirable?

(b) If so, what is its precise purpose? What should it contain? (For example: Just a programme of work to be covered? A statement of aims and methods?) For whose use is it? Who compiles it? Is it mandatory or a guide of suggestions? May it be varied (or even ignored?) by an individual teacher?

HOW TO INVOLVE THE PUPILS

(a) Is this by the use of textbooks? If so, in any particular ways?

(b) Is this by the use of 'assignments'? What different purposes might an assignment system serve? What might constitute the difference between 'better' and 'less effective' assignments? What advantages might an assignment system have over the use of textbooks? Are these alternatives mutually exclusive?

11 The Mathematics for the Majority Continuation Project

The Mathematics for the Majority Continuation Project (M.M.C.P.), as we have seen, grew out of the Mathematics for the Majority Project and work began in May 1971 with the intention of producing material for pupils of 'up to average ability' in the 13–16 age range. Originally this was to have been closely related to the *Guides* of the Mathematics for the Majority Project – but it was decided to extend the work to provide pupil materials based on a range of environmental topics. The first two packs, *Buildings* and *Communication*, were published in 1974 – in all, twelve such packs are envisaged, with the titles:

1 *Buildings*
2 *Communication*
3 *Travel*
3 *Physical Recreation*
5 *Seas and Rivers*
6 *Family*
7 *Motor Car*
8 *Countryside*
9 *Factories*
10 *Food, Shops, and Advertising*
11 *Fashion and Design*
12 *Holidays, Entertainment, and Recreation*

Some items from the *Buildings* pack are reproduced on p. 89.

Approach and rationale

One of the first statements of the aims of the project appeared in its newsletter:

> We would like to enable these youngsters to have some degree of control over their own lives and environment by extending their experience and powers of thought. We hope to help them towards understanding the structure and interconnections of a complex industrial society and will help them to use simple mathematical models as a means of achieving this understanding. We want to convey to the pupils the ways in which mathematical thinking is relevant to their other studies and convince them that mathematics is a tool worth using in problem situations outside the school. We intend the materials to be an agent for change toward a realistic form of education for non-academic children. (*Newsmaths,* **2** (1971), 112)

The project used a thematic approach in which the mathematics was quite deliberately subordinated to pupil motivation.

> The intention, in the first instance, has been to provide materials for use in mathematics classes which are not examination orientated. These materials are concerned with mathematics but are deliberately multivalent so that they can be easily absorbed into integrated studies and environmental studies courses. It is also possible to use these materials part-time with a more formal mathematics course or in any way the teacher may choose . . . We have tried to identify some of the objectives we feel to be important in regard to the pupil rather than the material (and the teacher). It would be wrong to believe that the objectives of mathematics teaching need to be defined in terms of mathematics or necessarily in mathematical language. For all but a very few people, maths is a means to an end. (Maybe a job, maybe entry into a course of F.E.). The actual attainment level is therefore very much the personal property of the individual concerned. So that a statement of objectives in terms of attainment level would be futile. (*Newsmaths,* **6/7** (1973), 8 and 9)

Each of the topics chosen has an environmental theme which is likely to appeal to the pupils in a variety of ways, many not necessarily mathematical. For example, one of the activities in *Buildings* leads to a discussion on how a fuse works and to ideas of safety in the home. Where such ideas occur naturally in context it seems desirable to follow them up. The subdivision of the curriculum into 'subjects' is after all a matter of convenience and any attempt at demarcation in this instance would be artificial – the teacher may well be only too glad of response and interest in *any* aspect of the work. But some teachers have seen dangers in this philosophy, arguing that there is a danger of so diluting the mathematical content that it disappears almost entirely. Considerable debate took place among teachers involved in the production of material, some of it reported in *Newsmaths*. Members of the Abbey Wood writing group wrote:

Extracting mathematics from the environment is bad enough, but even worse, we found, for less able pupils ... Too many bits of environmental mathematics are either too trivial or beyond the scope of the pupils ... It is a question really of what mathematics is, and the project is in danger of giving mathematics teachers, and especially non-mathematical teachers of mathematics, a false view. (*Newsmaths*, 1972, p. 3)

In an article in a later newsletter, from which we have already quoted above, the project director states the view:

It is not desirable to separate objectives for teaching mathematics from those general educational objectives which form the basis and justification for the existence of the school. ... We know from experience that mathematics education can and should contribute to (the intellectual growth of the pupil) but if it is failing in this task then its value comes very much into question, certainly if it is taken as an isolated study. (*Newsmaths*, **6/7** (1973), 9)

We are thus abruptly brought up against deep and difficult questions: the contribution of mathematics to the curriculum, the nature of mathematics, the basic rationale of the whole educational process. Most earlier projects had been able in the main to ignore these issues, since they could operate within a taken-for-granted set of assumptions about the nature and purpose of mathematical education. But, as we have already remarked in discussing the Mathematics for the Majority Project, these questions demanded consideration. In Kaner's words:

The (Mathematics for the Majority) Project was tackling one of the most difficult and frustrating phenomena within the educational system, one that had been swept under the carpet more or less since the creation of the secon-dary modern schools in 1944. (ibid., p. 7)

Materials

The project material appeared ultimately in the form of activity packs, boxes some $48 \times 30 \times 16$ cm containing about fifteen wallets, each dealing with a subtopic of the theme. Within the wallets are several assignments in the form of work cards and booklets, and sup-porting material such as games, information cards, and sometimes a tape recording (see figure 11.1). In some cases additional material needs to be provided by the teacher – such items are indicated on the wallets. A *Teachers' Handbook* in each pack gives notes on mathematical content, suggestions for use, and answers to problems. At £25 (December 1975) the packs are not cheap – it is estimated that about six packs cover about a full year's work. Some of the con-sumable (or 'losable') items are available as replacements; at this stage it is difficult to estimate durability or running costs. Perhaps a fair cost comparison would be with several sets of textbooks (which are becoming increasingly expensive) – but the high initial cost makes it less likely that individual schools will take the plunge by ordering and using these materials. In some cases local authorities have bought bulk supplies for distribution to schools, perhaps on a temporary or trial basis, which has made for wider availability.

The materials were devised by working groups of teachers, in most cases released from school one half-day per week for a term. Eight or ten schools might be represented in a group; each prepared classroom material in draft form based on one topic, which was tried out informally in the local area. The material was then sent to project headquarters at Exeter where a trial version was prepared, much of the design work being carried out by students in colleges of art. (It was considered that both the pupils' response and the quality of their work would be influenced for the better if high standards of presenta-tion were maintained, and by such means it was possible to achieve this even at the trial stage). The work of the groups was 'supported' by a team of nine regional coordinators, and the project team of five

You will need five more pieces of paper
about 12 cm by 20 cm

Fold them into box sections with sides of 2 cm

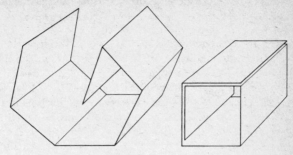

Support the box sections on the books
and load with washers, recording your
results, first on a table then on a graph as
before (you will have to support this
paper girder to keep it upright)

Collect all your results on
to a table like the one below

	Span (cm)				
Shape	10	12	14	16	18
⟨⟨⟨					
⊔					
▭					

Which shape gives the greatest strength?

Can you explain why ⊔ or ▭

should be stronger than ⟨⟨ ?

Figure 11.1 The last part of an assignment *Strength and Shape* from the M.M.C.P. *Buildings pack* (Schools Council Publications, 1974 – Schofield and Sims Ltd): a sheet of paper 12 cm by 20 cm is folded lengthways and used to make a 'bridge' between two books. The greatest number of metal washers which can be placed on the mid-point of the bridge without causing it to collapse is found. The experiment is repeated for different spans and for different ways of folding the paper.

at Exeter. The importance of involving teachers in the production of materials was thus re-affirmed and a noteworthy feature was the successful evolution of methods of organising this participation on a large scale – in all there were over forty working groups.

It was clear from the outset of M.M.C.P. that only classroom teachers were in a position to construct suitable materials for a large proportion of the pupil population. The ideas and assignments to be found in the wallets were originated by a large number of working groups in local education authorities in England, Wales, and Northern Ireland. Each of these working groups undertook to write assignments about an environmental topic and its interaction with mathematics. The choice of topic was agreed with each group in such a way as to avoid repetition of ideas and to maximise the total contribution. The result of this system of working is a collection of ideas far richer than could ever have been made by an individual teacher or even by a single local group of teachers working together. There is no doubt that there is still further opportunity for development in almost every pack – it would be an advantage to develop the more local aspects of each topic and no doubt many teachers will wish to introduce further ideas arising from their knowledge of their own pupils. The physical design of the materials is intended to facilitate this local development.

During the writing period the group members exchanged ideas amongst themselves so that any collection of material which was submitted to the project had, to some extent, already been presented in the classroom and had its weaker points improved. It has been evident that the participation of these large numbers of teachers has been valuable in preserving a clear view of the realities of the school situation and the pupils' abilities. (Publisher's advertising material, Schofield and Sims Ltd)

It has been in the construction of materials that the project has had its most outstanding success. The strategy of involving teachers as writing with a limited task combined with the support of local education authorities in releasing teachers to undertake this work, has led to a richer set of materials than could possibly have been imagined in the beginning. This has been partly because the teachers were asked to explore mathematical possibilities in areas which were both unfamiliar and interesting to them. There is a vivid contrast between the contributions made by teachers who are thinking about, for example, the sea to those made when a group of mathematics teachers are asked to find new ways of presenting 'ratio and proportion' to children. We feel that the project strategy released a surge of creative energy that had previously been untapped in any sort of curriculum development. (*Newsmaths*, **8** (1974), 5)

Examples of environmental 'starting points' were given in *Newsmaths*. Thus issue No 3 contained suggestions for relating mathematical ideas to football (perhaps building on intuitive ideas which children already possessed); goalkeeping, (where to stand, does it help to be tall?); marking a player (one to one correspondence, permutations). 'Work' was a theme of many suggested activities: statistics (pupils' part-time jobs, types of job available locally, hours worked, wage rates); types of mathematics used in a given job; heart-rate after spells of physical work. No 5 was a special calendar issue (motion of the earth and moon; longitude, G.M.T., and sun-time; making a sundial; the French Revolutionary calendar). (See also Pass 1976.)

The emphasis on group work and on pupil activity was a re-iteration of one of the theses of the parent Mathematics for the Majority Project, and there was a deliberate attempt to encourage as much variety as possible in the form of materials produced. Inevitably, however, cost factors and other practicalities tended mainly to restrict the range to the conventional work card and booklet. The use of photographs and graphics, while making the material attractive to pupils, also tended to push up the cost.

As the quotation above indicates, it is envisaged that some of the material will be adapted by teachers to suit local circumstances; indeed the more an item, originally developed in one locality, makes use of factors of local interest, the more compelling will be the need for such modification. A minor problem arises now for some teachers who may feel that their home-made materials compare unfavourably with the more professional appearance of the project items, though this is reduced by the availability in some teachers' centres of fairly sophisticated reprographic facilities.

Trials and evaluation

Trials of the materials were arranged so that both the schools

represented in the working groups and a few others not otherwise
connected with the project were involved, about fifteen to twenty
schools for each pack. (No writing group was involved in the subsequent tests of its own products and no school tested more than
three of the twelve trial packs.) Almost from the beginning the
project had its own evaluator, and later the evaluation team was increased to three. Questionnaire forms accompanied each trial pack as
it was sent out; the evaluators visited the schools, interviewing
teachers and children for their comments on the material. Feedback
reports were then produced for use in re-drafting each pack in
preparation for publication. There were inevitably some delays in the
return of questionnaires, but, perhaps because of a realisation of the
importance of prompt collection of information if it was to be effective in the re-writing, and helped by the availability of extra staff at
the crucial stage, the evaluation data seem to have been effectively
used.

Generally the packages have been very well received in the schools and the
pupils enjoy using them. It has been rewarding to see pupils somewhat unwilling to stop work at the end of a lesson and to discover that a number of
pupils have asked their teachers for some revisional work in mathematics to
enable them to work the cards — 'Please Sir, can we do some work on
percentages?' (*Newsmaths*, **6/7** (1973), 34)

The role of the evaluator in such situations is a difficult one. Though
he (or she) is a member of the team, he must not become so integrated
with its activities that he is unable to be objectively critical in his
assessments — nor must he allow antagonism to develop between
himself and the other team members. A delicate situation! '. . . at all
times I have tried to act as a friendly critic rather than a devil's advocate' wrote G. W. Manfield.

We have to face the fact that in education, all evidence is to a large extent
subjective evidence. Even those enquiries carried out by the impersonal
means of questionnaire or testing are the result of a number of previous subjective decisions on what questions to ask, what analysis to make . . . So
when we, as a project, look for evidence about our own work we try to find
unlike people who corroborate one another's observations. The greater
diversity among observers, the greater likelihood that they are near to an ac-

curate description of the phenomena that interests us. Best of all amongst the observers are the pupils themselves and it is from their response and opinions that we judge mainly whether a piece of material is worth including in the final publications. Thus we use as evaluation corroborated subjective statements from a wide range of pupils, teachers, and others, statements which give more than just an intuitive idea of the things which please and excite the pupils, which animate their nature and which arouse their (by now) dormant learning potential and mental powers. Thus we proceed by intuition based on a complex mixture of awareness and ignorance, a pragmatic system of development plus evaluation. (*Newsmaths*, **6/7** (1973), 9)

An interesting independent exercise in evaluation was carried out by Mr W. Flemming and reported in *Newsmaths*, **8** (1974), 8. Students in initial teacher training collaborated with local teachers in an examination of the use of the *Travel* packs (trial version); they tried to assess how far the purposes of the project were being achieved through the use of these materials and made several practical suggestions for improvement (including the idea of a set of '*Notes of guidance*' to help teachers to become familiar with a newly-acquired pack). Children's attitudes were measured and there seemed to be clear evidence that the *Travel* pack was more favourably regarded than was 'normal' mathematics, at least by these children.

It was noted that the trial packs were used in a variety of ways, some teachers complementing the work with one or two normal lessons per week on basic work, or to cover common points of difficulty which came to light while the packs were in use. Since, in general, 'it was found best for pupils within a class to work on different items', the published versions contain only one example of each.

Information was also collected by the project team on the resources of mathematics departments, on the organisation of remedial work in schools, and on the effectiveness of the working of writing groups; all issues of considerable relevance to the work of the project, though arguably the information might well have been more valuable if collected at an earlier stage in the life of the project.

Communication and dissemination

The Mathematics for the Majority Project had established a network of communication with pilot and associated schools. For the continuation project it was even more vital that the contacts already made should be exploited and that links with the teachers' writing groups should be strong; the project seems to have been remarkably successful in achieving this. The small central staff, working with the regional coordinators, was used to orchestrate and support the writing efforts by teachers in the field and, by concentrating energies on the participating teachers' groups, was able to maintain contact and ensure that the work of writing went forward. In some areas groups of teachers already existed which had been involved in the parent project, others were associated with teachers centres or brought together by L.E.A. advisers or college tutors. There were difficulties over half-day release (though there would have been more a few years later!). The writing itself was not easy:

I never realised:
... how much time it would take
... how long it would take to arrive at clear, fundamental aims
... how essential it was for each member of the group to have a plan of his own activities and know his place within the structure of the group
... how many novel ideas can be generated by a small group
... how helpful some firms and authorities can be
... how insecure some teachers can feel when on unfamiliar ground
... how difficult it is to keep to a one-term timetable '
... how enjoyable working with a group of like-minded teachers can be
... how little I know about rivers, climbing, and games. (N. A. Pass in *Newsmaths*, **2** (1971), 6)

Yet the process was profitable:

It has been said again and again by the leaders of the writing groups that even if no material had been forthcoming, the process would have been well worthwhile for those who took part. (*Newsmaths*, **8** (1974), 4)

Project news was made available in *Newsmaths* which appeared first in summer 1971 and thereafter three times per year. The parent project *News Sheet* had been produced on newsprint, and was issued free at irregular intervals. By contrast *Newsmaths* was printed on good quality paper, with plenty of photographs and specimen 'pages' from forthcoming packs, and sold at 20p per copy. It may have been a more effective means of communication than its predecessor, though probably reaching fewer people. Perhaps in the same spirit of concentrating resources, 'publicity' conferences were only arranged at a fairly late stage in the life of the project. (The overlap with the parent project had meant that the problems of establishing initial contact with interested teachers were greatly eased.) The final issue of *Newsmaths* (**8** (1974)) contained a list of 'regional contacts' – teachers and advisers who had dealt with the materials and would be able to answer queries, etc.

... (the publicity conferences) ... are concerned with the setting up of support structures which would outlive the project. When our materials are published and are available for general use, there must exist some organisation within regions which can organise training conferences and give support direct to teachers in schools working with new material. (*Newsmaths*, **6/7** (1973), 34)

A short film was also made, describing the organisation of the project, the operation of the teachers' groups, the style of material produced, and how it was being used in classrooms. This provided an alternative to the traditional presentation by lecture, an alternative which was perhaps more consonant with the style of the project and had the merit of being readily transportable.

Problems arose in the final stages as the team began to disperse; the director himself left the project shortly before its termination. N. A. Pass, who had been associated with the work as a regional coordinator, succeeded him to oversee the final re-drafting of the mass of material already prepared for the remaining packs, so that these could go to the publishers, and to encourage the mounting, by local education authorities, institutes of education and any other interested bodies, of courses which would make the material as widely known to teachers as was practicable.

Evidently some lessons had been learned from earlier experience – dissemination was now accepted as necessary but it is arguable that, even so, the resources allocated to this task and the final editing were not enough, and that the new director was given a near impossible brief.

In a sense the materials must speak for themselves – and it is not suggested that a Schools Council project should carry an imprimatur implying any limitation upon the freedom of teachers and schools to accept or reject its wares. Nevertheless, unless the product is made accessible to teachers, its purposes elucidated, and the experience of its use made available, it will not be possible for anyone to make informed judgments about it, and there is the tragic possibility of much potentially fruitful effort being partially wasted. This is not an easy matter. The establishment of permanent Schools Council Dissemination Units might create more problems than it would solve, nor can projects be funded into an indefinite future at full strength. In any case it is unlikely that high calibre project staff would be willing to remain for long periods; such work may well be seen as a temporary phase in one's career and the project's timing may not fit particularly well with job opportunities for its senior staff. (It is said, cynically, that a project director has three major tasks – to apply for an extension of the project, to carry out the project itself, and to tidy up and look for a new appointment!)

Some issues raised

The Mathematics for the Majority Continuation Project demonstrated that teachers' groups working locally with half-day release and receiving outside support can produce useful curriculum material, together with considerable, though less obvious, advantages in terms of teacher motivation and in-service training. By giving central support to grass-roots teacher-based curriculum development, the project helped to shift the emphasis away from the earlier model of centrally-initiated curriculum change. This, however, is still a live issue, and it is not self-evident that a similar pattern will or should be followed in future developments.

The project has made a considerable impact on the problems of teaching mathematics to the less able, but though a lot has been achieved, many unresolved questions remain. Was the Mathematics for the Majority Continuation Project, as some asserted, really an environmental studies project, containing very little mathematics? And if so, is this necessarily a bad thing?

What were we as mathematics teachers really trying to do for our weaker pupils? . . . Is mathematics really important for them? . . . What ought we to be doing for them? And how best can we do it? (*Newsmaths*, **8** (1974), 4)

12 The Schools Council Sixth Form Project

Rationale and early stages

The Schools Council set up the sixth form project in January 1969, although the main phase of activities started in January 1971. Initially a 'Study of various aspects of sixth form mathematics' was commissioned and C. P. Ormell, who was appointed to conduct this, was given a mandate to research in four areas:

1 A general review of the content of sixth form mathematics, aimed at assessing the value of various topics, and perhaps especially those of recent introduction, for different categories of pupils

2 An investigation into the connection between mathematics and physics. This would consider not only the range of mathematics needed to facilitate the effective teaching of physics, but also whether some topics, notably some of those conventionally included in applied mathematics, would not be better regarded as part of physics

3 An investigation of the mathematical needs of pupils taking other subjects such as biology and economics

4 An investigation into the mathematics that can profitably be taught to the non-mathematician (i.e. the pupils whose main interests do not include mathematics or subjects requiring mathematical servicing) in the sixth form. (*Newsletter*, **1** (1969))

In this initial phase he undertook an extensive analysis of the situation and of the possible approaches, producing a number of Discussion Papers. In these he considered the mathematics/physics interface (attempting to define the problem of demarcation and mutual support in the relationship of sixth form mathematics and physics) and the analysis of syllabus content. Ormell also produced some draft curriculum packages (on indices, quadratic models, and limits) intended for trial in schools, which illustrated the approach which he proposed for the project. Two aspects which were elaborated in articles published later by Ormell are those of relevance and applicability.

Ormell's two varieties of relevance, 'horizontal' and 'vertical' have already been mentioned. He points out that the 'deferred reward' associated with vertical relevance ought not to be too long delayed, yet in many instances the justification for a piece of work is long postponed, or even never reached (as with pupils who will not ever pursue mathematics to the required level). This is an important issue – all who have studied mathematics will readily identify topics which for them developed meaning and pertinence only from a later vantage point.

Vertical relevance is conveyed to the student by the teacher; and this is, as it were, a promise of visible (horizontal) relevance to come. The important point is that these promises should be fulfilled, and be seen to be fulfilled, by later work. (*Newsletter*, **4** (1971), 30)

It is arguable that in the nature of mathematics, some approaches will have to be taken on trust if the teaching is not to become impossibly long-winded. Ormell's contention is that this situation should be avoided wherever possible and this can be achieved in large measure by concentrating on the view of mathematics as 'applicable', and the 'science of possibility':

The essence of the idea of hypothetical relevance is that one can use mathematics to explore the possibilities of newly postulated situations. (*Newsletter*, **2** (1970), 18; see also Ormell 1972a, 1972b)

Mathematics was seen by Ormell not as the study of structure but as the exploration of the possibilities of real situations.

The implied criterion in many schemes is not that children should be led to think about the possibilities of real situations (though this may be included as a minor theme among others) but that they should be led to think in terms of the hierarchies of formal concepts needed in modern academic university pure mathematics. It seems unlikely that such thinking will be profitable for any but a minority of our students. To base the mathematical education of the many on the needs of the tiny minority who will eventually become professional pure mathematicians is difficult to justify indeed. But simply to remark this is not enough. What we need is a valid criterion of 'syllabus worthiness' on which to base the mathematical education of the many. (*Newsletter*, **4** (1971), 28)

The relevance of mathematics then lies in its use as a tool in providing models for investigating possibilities, and the situations which are 'modelled' provide both the context and the motivation for the topics studied. Ormell does not underestimate the difficulty of setting up and using a mathematical model, but suggests that this activity must be central to the course, not just because it provides motivation, but because this, for him, is 'what mathematics is all about'.

A view sometimes expressed is that the application of a piece of mathematics to a situation is an activity of a lower level than that of studying the same piece of mathematics *in vacuo* . . . a straightforward affair, calling for less ability and less intensity of concentration than that of studying pure mathematics . . . discussing applications should be an important component in the mathematics syllabus at every stage . . . We have tended systematically to underestimate the judgement and understanding needed to apply mathematical models appropriately and usefully to real situations.

If we want mathematics to make sense to the mass of intelligent students who wish to study it, we must ensure that they see that it has a use (the use is essentially illuminative), i.e. that it has a social purpose over and above that of providing a few exceptionally talented people with a unique aesthetic experience . . . The thing we can do best with mathematics is of course investigating possibilities. Science at the growing point consists largely in forming hypotheses, i.e. forming possible explanations of a situation, and then investigating their consequences . . . So here we have the basis of a new way of motivating mathematics. What mathematics can do best is urgently needed on a massive scale in science, technology, and social progress. (*Newsletter*, **4** (1971), 32 and **6** (1973), 41)

The main phase

When the main phase of the project began in January 1971, extra staff were appointed and work began on writing 'relevance-enriched' material for trial. These included introductory books for pupils and teachers (necessary to explain the purposes of the project since the approach, essentially process- rather than content-orientated, was likely to be very different from anything either pupil or teacher had encountered before, and might easily be misunderstood). Packages covered topics such as binomials, geometric series, linear models, generalisation, and limits. J. B. Morgan (as editorial adviser) and two research associates joined the project; C. Bentley became evaluator and trials manager. 102 schools 'widely scattered over England and Wales' were used in this stage of evaluation and a Workshop Conference was held in September 1971 for teachers from the trial schools 'to give them an opportunity of becoming familiar with the aims of the project, the material of the packages and methods of evaluation.'

(The project's) principal tasks are to assess the value of the topics in existing sixth form syllabuses, both for mathematically-gifted students and for others differently motivated, and to study the needs of user subjects such as physics, biology, and economics. The trials of the project material are designed to collect information and advice about what could be called the 'New Model' view of mathematics. (J. B. Morgan writing in the *Mathematical Association Newsletter*, **21** (1971))

Information about the project was given in the steady stream of duplicated 'project' papers and 'discussion' papers, obtainable on request from the project office, or in a duplicated *Newsletter* (issued twice a year for a small charge — those from No 6 onwards were called *Polymetrics*), in *Trial News* (brief progress reports and announcements), or in *Reading Area News* (an early newsletter from local teachers). The first published texts appeared in 1975 (see p. 108).

To follow-up the idea of modelling a 'real' situation, the students were encouraged to submit projects — called 'feasibility studies' — in which mathematical ideas were applied to a design proposal for an innovation, e.g. car park design, design of a bus-layby, of an accelerator for stepping on to a fast-moving pavement. Several papers were published illustrating this idea: *Doing a feasibility study*; *Local feasibility study* No 1 (the Reading cable car system); *Brockenhurst College feasibility study* (an account of pupils' work, extending over fourteen periods, in designing a passenger transport system linking the college and the railway station, with comments from the teacher and a project team member). Other forms of suggested course-work were:

1 Setting up progressively more accurate models of a situation e.g. jostling in a school corridor
2 Investigating a real (as opposed to hypothetical) object, e.g. a bicycle
3 Providing instances of a particular type of model, e.g. logarithmic, indicial, linear

All this course work formed part of the assessment in an examination, at G.C.E. AO level, in basic applicable mathematics, which was staged in a pre-pilot version in 1974 and as a pilot examination (conducted by the University of London School Examinations Council on behalf of all G.C.E. boards) in 1975 (see p. 109). A specimen examination paper was produced in which an extended series of questions were asked, related to a specific physical situation (the increased vertical force on train wheels, leading to increased braking force, consequent on passengers' rising from their seats as a train approaches a station). The questions were quite demanding, both in the insight into the application of the mathematical model demanded (which *was* under test) and in their interdependence (which would have meant that candidates who were unable to answer questions at a specific stage would probably have been unable to answer later questions because of this). This difficulty was circumvented by the device of providing sealed 'hints', which gave candidates the opportunity to progress beyond the point of difficulty – though by using (unsealing) any hints they incurred a 'penalty', losing some credit in the process.

Several interesting questions arose, which were investigated in the pre-pilot run of the examinations. Would the questions be altogether too difficult? (In the nature of the trials, students generally spent only a few weeks using the project materials, so as not to disrupt their other work. Consequently they had had little opportunity of exposure to the way of thinking now demanded of them and little practice in using it.) Would students use the hints unnecessarily to 'check' their answers through feelings of insecurity? Would use of hints be related to ability? What was the appropriate penalty for use of a hint? This was an interesting attempt to see whether this type of question could be asked as a part of normal examination procedures for pupils in general. (There are examinations, such as those for University Entrance Scholarships and the Russian Olympiads, where exceedingly difficult questions are set, in which only outstanding candidates can answer more than a few questions – but these are aimed at a different target and serve a different purpose.) The use of assessed course work – now becoming more common with a growth of Mode III C.S.E. – may prove to be a better way of tackling the problem. Only after further experiment, however, can such a question be answered, and this was a good example of an attempt to suit the assessment to the course, rather than vice versa as all too often is the case.

The problems of project evaluation also received considerable attention. Ormell contributed a chapter to the Schools Council (1973) publication and several project papers dealt with this topic. In the main the accent was on formative evaluation (obtaining feedback on the effectiveness of the materials in furthering the goals of the project). Peculiarly difficult problems were faced here. As the approach was so radically different from that to which most teachers and pupils were accustomed, it was not even clear initially whether the goals were realistic or not. Thus the method adopted was to try out the materials in the belief that the task *was* possible, provided suitable 'packages' could be constructed.

Ideally, to give a fair trial to the materials and philosophy of the project, it would have been desirable to provide a whole A level course, or something approaching it, on this basis. (Such a radical change of viewpoint is not rapidly assimilated and it was *a priori* likely that only an extended exposure to the ideas and methods would have an effective influence on pupils' ways of thinking and abilities in this way of working.) Unfortunately this was largely impracticable. It might be thought that students embarking on A level courses would be put at risk if they were an experimental group in such a course. No appropriate examination existed – one would have to be constructed from scratch. This was a nettle which S.M.P. had grasped, by creating their own S.M P. A level examination – but only after extensive consultations with universities as to its acceptability, and intense discussion, with some internal dissention, on its format. This was a development from within; propelled by the enthusiasm of teachers in the participatory S.M.P. schools, many of them independent, with long traditions of examination success, and resulting confidence to launch out into the unknown. Many of the schools prepared to innovate were already committed to S.M.P. or M.E.I. A level projects and it was not to be expected that such a radical course of action would find widespread support among other schools. On the other hand, pupils who were studying mathematics in the sixth form and not taking A levels would, in general, be expected to be neither so proficient nor so well motivated as those following a two-year G.C.E. course. Thus extended trials were almost ruled out, and the only way forward was the piece-meal evaluation of each item in short-exposure trials. In a few cases it was possible to arrange extended use of project materials, and some interesting results were reported (e.g. J. Coulson in *Polymetrics* 6 (1973) writes of the varied response of her students at Luton Sixth Form College. One 'complained' of having to think while using the packages; others were held back initially by a feeling of insecurity in the more open situations. The individual work which the packages necessitated allowed the teacher to give increased attention to single pupils and so overcome some of these problems. Success was not uniform, however; one pupil complained 'It's work, whatever you have to do!') Although an evaluator was designated, he also contributed to the writing of materials and the evaluation seems to some extent to have been shared between members of the team, rather than being conducted in the 'classical' independent evaluator model.

The materials so far produced (December 1975) have several features of interest (see figures 12.1 and 12.2). They (will) comprise both pupil and teacher books — since both parties are likely to need some initiation into the approach — with supplementary background material and further problems. Within the pupils' books, mathematical and background references are given, together with hints and answers whose numbering is scrambled (so that they are not likely to be read accidentally in advance before students have thought about the question for themselves). An early project paper (*Semi-programming package material* (1970)) discussed the extent to which the students need be led along a pre-defined path through such material; the hints provide a means of allowing flexibility while ensuring that students are not 'lost' *en route*. An Expert Advisory Panel was also set up (in December 1972, with thirty-two members) to provide advice on the validity and authenticity of the examples and background information in the project materials. This also acted as a source of further ideas and variations of modelling situations.

The first part is a set of questions concerned with techniques developed in the text, and is designed to help you in the handling of the mathematics in the problems.

The second part tests interpretation of formulae.

The questions in the third part describe situations which we can investigate using the logarithmic function.

Numerical answers should be given to three significant figures unless otherwise stated.

Techniques

1 Integrate:

$$-x^{-1}, \quad \frac{x-1}{x}, \quad 12/x, \quad \frac{2x^2+1}{x^3} \qquad \textbf{A 341 \quad H 522}$$

2 Find dy/dx if $y = \log_e(0.1x)$. What is the gradient of the curve
$$y = \log_e(0.1x) \text{ when } x = 4? \qquad \textbf{A 500 \quad H 756}$$

Figure 12.1   Extract 1 from Schools Council Mathematics Applicable,
Logarithmic/Exponential (Heinemann, 1957, p. 82)

Interpretation...

4 A security firm offers combination locks in which the number N of possible permutations ranges from about 1 million to about a billion billion (10^{24}). **B 123**

Why is it not convenient to show the cost of locks of various values of N directly on a graph? **A 586**

If the 'security' of a lock is defined as $\log_{10} N$ and the cost, C, of a lock of security $\log_{10} N$ is £2.5 $\log_{10} N$, find dC/dN and hence dN/dC.
A 622 H 778

At which end of the scale does one get the greatest increase in N (δN) for a given small increase in C (δC)? **A 484 H 705**

Draw sketches of N against C and $\log_{10} N$ against C for locks ranging from zero security to a security of 24. **A 751**

Is it true that for locks costing £10 or more an additional cost of one penny would, in theory, increase N by more than 100? **A 674 H 546**

Problems

1 The rate at which a certain new woollen sweater (initially bone dry) will absorb moisture after a time t (seconds) when worn in *very humid* conditions is said to be k/t (mg/s). **B 124**

Assuming this to be correct, what does

$$\frac{k}{t} \, \delta t \qquad \qquad \textbf{A 722}$$

represent?
What does

$$\int_{t=1}^{t=60} \frac{k}{t} \, dt$$

represent? **A 526**

Evaluate this in terms of k. **A 649**

Is it true that the sweater will absorb as much moisture in the second second as in the second minute? **A 757 H 579**

Find k (to the nearest whole number) if the sweater absorbs 350 mg of water in the second second. **A 785 H 498**

Figure 12.2 Extract 2 from Schools Council Mathematics Applicable, *Logarithmic/ Exponential* (Heinemann, 1975, p. 85). The symbols **A**, **B**, **H** relate to answers, background notes and hints.

Comment

Though it is too early to say what its impact on the schools will be, there is no doubt that this project has examined some of the major issues in teaching sixth form mathematics, several of which had received little attention in earlier developments. The first wave of mathematics curriculum reform was conducted in almost complete isolation from the (Nuffield) developments in the sciences. Ormell's attempt to provide a solution to the 'maths/physics problem' by regarding mathematics as a model-building activity (which uses a good deal of what may be thought of as 'traditional' mathematics) seems more helpful in this regard than the navel-contemplation of a preoccupation with structure, and though to devise a mathematics syllabus to satisfy all client subjects is almost certainly beyond the wit of man, his wide choice of examples from technological and biological fields does provide good opportunities to forge links if the will is there.

The related issues of relevance and applicability, too, though sometimes presented with a flavour perhaps too philosophical for many, are important issues squarely faced; there are many complaints that students in sixth forms 'study mathematics blindly ... suspend the critical faculties ... accept unquestioningly the concepts and techniques presented ... conform to a pattern which (they hope) will maximise the expectation of obtaining good qualifications (Ormell 1972a, p. 126).

If these materials are adopted in a widespread fashion, complaints may still arise in future – but they are likely to be different ones! It is difficult to envisage this sort of material being taught as a cram course. The examination/assessment problem may well turn out to be crucial and here the project has made imaginative attempts to move forward into new territory. (We shall see in the next chapter that a similar problem faced those who attempted to develop this theme of 'mathematical investigations'; a successful outcome on the examination front was a pre-requisite to progress there also.)

The AO examination of G.C.E. has not, in general, a large following, though proposed and impending changes in the examination structure – the Certificate of Extended Education and the suggested normal (N) and Further (F) levels at eighteen plus may give a more fluid situation with possibilities for this project to gain a foothold.

The seven exploratory studies on N and F examinations in mathematics commissioned by the Schools Council include one to be conducted by the Sixth Form Project. (See Quadling 1975.)

The earlier developments at this level have to some extent already monopolised the field. It remains to be seen whether the materials, now ostensibly intended for 'a general post O level course of mathematics suitable for non-specialist students in colleges and sixth forms', will prove sufficiently attractive to be used more widely and so infiltrate the courses for specialists also. Much will depend on the resources which are available for making the materials known and disseminating the experience accumulated by the team and the trial schools.

Appendix

1 **Titles of 'Mathematics Applicable'**
   (Publications of the Schools Council Sixth Form Mathematics Project; publisher: Heinemann Educational Books)
   Mathematics Changes Gear (students' introductory book)
   Presenting Mathematics from the Applicable Point of View (teachers' introductory book)

   (a) Students' starter units
   Understanding Indices
   Geometry from Coordinates
   Introductory Probability
   Polynomial Models
   Vector Models
   (b) Students' continuation units
   Expanding Binomials
   Taking Limits
   Logarithmic/Exponential

Sequences and Series
Calculus Applicable
Introductory Dynamics
Probability and Statics
(plus teachers' units corresponding to each student unit)
(c) Supplementary material
Problems in Applicable Mathematics (parts A–L)
Advanced Background Book (parts A–L)

2 Some articles published by project staff

KNOWLES, F. L. 'An Approach to Applicable Mathematics', *Mathematics Teaching*, **56,** Autumn 1971, pp. 50–3.

ORMELL, C. P., 'Newtonian Mechanics and the Sixth Form Syllabus', *International Journal of Mathematical Education in Science and Technology*, **2,** 1971, pp. 233–41.

ORMELL, C. P., 'Mathematics through the Imagination', *Dialogue* No 9, Autumn 1971, pp. 10–11.

See also Ormell 1972a, 1972b and Schools Council 1973.

3 Papers relating to the proposed project examinations

(a) *Pre-Pilot Examination Course* 1973–4 (pupils successful in the pre-pilot examination received a 'letter of attestation' from the project). Paper 15/73.

(b) *A Pilot Examination 1975–6. Analysis and Development.* Paper 16/73.

(c) *Pilot Examination Syllabus.* Paper 19/74. The pilot examination at AO level was a 'special-syllabus' examination of the London Board, and restricted to 'official' pilot schools (135 candidates from 16 schools). Marks were allocated as follows:

Course work: essay or 'feasibility' study – 20% (See *Course Work*, project paper 17/73)
Paper I (1 hr): standard techniques – 20%
Paper II (1 hr): abstraction and interpretation skills – 20%
Paper III (2 hrs): a sustained discussion of a situation,

involving answers to twenty related questions, with additional hints (their use involving a mark penalty), check-ups (for reassurance – no penalty) and notes – 40%

The results are discussed in:

BENTLEY, C. and MALVERN, D. D., *Looking Back on the Applicable Mathematics Course 1975*, which also gives evaluation data on student reactions to the course.

The pilot AO level examination will be offered again in 1976, and in 1977 will be available on an unrestricted basis through the London Board on behalf of all G.C.E. boards. (See also: *Proposed AO Exam: questionnaire results*, Paper 12/72; *Proposed AO Exam: use of hints*, Paper 13/72).

13 The idea of mathematical activity: a development in some colleges of education

The approach

It is apparent that curriculum development does not occur solely in formal ways associated with national or regional projects. Indeed many would argue that the only worthwhile form of development is the informal activity resulting from the initiative of teachers themselves and supported, rather than stage-managed, by more centralised resource-providing bodies; attempts to graft 'official' schemes of curriculum reform on to local circumstances may lead to a form of 'tissue-rejection'. (In so far as they achieve success, however, such local initiatives become identified with particular groups or organisations, and there is the danger that they will then become another example of a formal, because identifiable, viewpoint.) Some recent developments associated with the idea of mathematical activity are an instance of this informal mode of curriculum reform; not only do they represent an interesting view of the nature of mathematics teaching, but they illustrate some of the difficulties and limitations surrounding all such development.

The theme of mathematical investigations was not new, as we have seen in chapter 2, but came into prominence particularly through the activities of members of the Association of Teachers of Mathematics. In 1966 a report was prepared by the Association as a contribution to the International Congress of Mathematicians held in Moscow (Association of Teachers of Mathematics 1966). This compilation of the work of many individuals was entitled *The Development of Mathematical Activity in Children; the Place of the Problem in this Development*; the ponderous title was probably necessary to convey the spirit in a way which the words 'problem-solving in mathematics' would not have done. Two items which outlined work at Nottingham College of Education (A. W. Bell) and Abbey Wood Comprehensive School (D. S. Fielker) were particularly significant as examples of what was being attempted. The underlying philosophy may perhaps be best summed up in an extract from a report by A.T.C.D.E.

In the training of a practising mathematician at any level, whether he is creating new mathematics, or new applications of old mathematics, or merely making routine applications of well-known ideas, we can distinguish two parts, the learning of a range of mathematical ideas and techniques, and the development of the ability to do mathematics, that is to use the ideas and techniques to solve an appropriate range of problems. The first of these parts seems normally to be undertaken in an educational institution, and the second 'on the job'. Typically, for a research mathematician, the first degree represents the first of these and his Ph.D. work the second; whereas a mathematician going into industry usually leaves the university with a first degree and completes his training in the industrial laboratory; and the routine user of mathematics leaves school with, for example, A level mathematics and then learns how to use this knowledge as required in his occupation.

Although it is true that the learning of mathematics has always been considered to involve the solving of problems, there is a total difference of approach between problems set as exercises on specific ideas which are being learned, of which the aim is to establish the ideas securely, and problems which exist in their own right, for their own intrinsic interest, and for the solving of which no particular method is specified. It is the exploration of these more open problems which we feel to be the essential characteristic of real mathematical activity. (Association of Teachers in Colleges and Departments of Education 1967, chapter 2, p. 3)

In one of his excellent series of books on the processes of mathematics, Polya reprinted his lecture to the Mathematical Association of America on 'Learning, Teaching, and Learning Teaching', in the course of which he said:

If the teacher has had no experience of creative work of some sort, how will he be able to inspire, to lead, to help, or even to recognise the creative ability of his students? A teacher who acquired whatever he knows in mathematics purely receptively can hardly promote the active learning of his students. A teacher who never had a bright idea in his life will probably reprimand a student who has one instead of encouraging him. (Polya 1965, p. 113)

Work in schools

The extent to which an approach based upon doing mathematics could be adopted – mathematics as a 'process' rather than the 'product' (of other men's minds) to be acquired and mastered – depended on the freedom of the institutions concerned to devise their own courses, and upon the examination requirements which were imposed upon them. In the school context the strategy adopted was to put forward a Mode III C.S.E., or school-based O level G.C.E., 'syllabus', individual schools negotiating an agreed form of assessment with the examining boards. (This is an option which has always been open to schools, even before the advent of C.S.E. Mode III. Admittedly the procedure can be long-drawn-out and is arguably loaded against the school, which has to justify its proposals.) Examples of some of the question papers which resulted are given in Association of Teachers of Mathematics 1968, pp. 65–7. Attempts were also made to set up an A.T.M. Sixth Form Project and a joint submission was made to a G.C.E. board by a number of schools where teachers had agreed to work together (Bell 1968). A.T.M. also produced several issues of a *Sixth Form Mathematics Bulletin*. The first appeared in May 1969 and others followed at six-monthly intervals. These reported progress in discussions with Schools Council and the examination boards, explained participants' approach to sixth form teaching, and gave ideas for investigations and examples of pupils' work. Several discussions and planning meetings preceded this submission – the following quotation from an informal discussion document outlines the approach.

The way we understand the nature of mathematics underlies the philosophy of the project. It could perhaps be said that, for us, the way of working we are developing is not an alternative way of teaching mathematics, but the only appropriate way, given our views about mathematics itself. ... mathematics is best understood as an activity and not as a body of knowledge or a set of formal structures (although of course mathematics 'has' knowledge and 'has' formal structures that have come about as the result of mathematical activity). But mathematics is 'done'; it is constructed and developed by man. 'One can make a piece of new mathematics in the same way that one can make a new pair of shoes.' It is a more fruitful question to ask 'how' mathematics happens, or 'when' mathematics arises, than to ask 'what' mathematics is. It follows that it is not very helpful to talk exclusively of 'understanding' mathematics, or to analyse it in conceptual terms. The operations and processes of mathematics, the dynamic of mathematical activity, are the things to be grasped, comprehended, and taught. (D. H. Wheeler, personal communication, 1968)

In the event, the Sixth Form Project did not materialise in the way which had been intended; a parallel proposal for an O level project was also abandoned (see Association of Teachers of Mathematics, 1971). Discussions with the Southern Universities Board led ultimately to agreement on an A level scheme, though this was available only to 'double mathematics' candidates. This condition ruled out several of the small number of schools originally involved, and others, as a result of staff changes, were unable to go forward, so that, sadly, this initiative came to nought. The difficulties were many, and were not confined to reaching agreement with the G.C.E. board on mutually satisfactory terms. Among the issues which had to be faced was that of 'syllabus coverage' – was it possible to specify content areas or mathematical techniques which it could be guaranteed would be covered by all students? Would a paper to test these minimal techniques be acceptable as part of the assessment? Was the work on 'investigations' to be assessed by a formal exam paper? Could this be adequately examined in a formal paper, or would such an attempt merely result in students being 'trained in ways to pass investigation papers'? At the heart of the problem lay the considerable investment of time required by the proposed approach to mathematics. This time could not be given under the current conditions if the syllabus were to be covered – consequently only a root-and-branch proposal, in which the guaranteed syllabus coverage (if any) was reduced to a minimum, would be workable. One teacher even suggested that the student's freedom of choice should be complete, so that he would 'set his own syllabus by hindsight at the end of the course'.

Problems abound also in the mechanism of assessment of work of this type. How much of any investigation is the student's own work, and to what extent should help from the teacher count against his performance? Yet what of the adverse effect on the student's learning if the assessment aspect should dominate the course in this undesirable way?

College of education 'main' courses

All the considerations referred to above applied equally in the colleges of education, though here other factors made the position easier. Whereas an O or A level result is often used as an entry requirement for some subsequent course, the college of education certificate course is, in general, a terminal course in its own right; provided that the college is satisfied that it adequately prepares the student for his role as a teacher, the 'users' of the qualification, schools and local education authorities, are more ready to take its validity on trust. The situation in the colleges of the Nottingham Institute in the early 1970s provides a case history in which the writer was personally involved as an examiner. (It was not unique, for similar developments were occurring elsewhere as a result of the interchange of ideas through A.T.C.D.E. conferences, etc.) It was suggested that the assessment be based on a selection of the work done by the candidate during the course, in some cases together with an oral examination, to take into account all aspects of a student's work and to include most or all of the following items:

1 Individual mathematical investigations
2 Exposition of material selected by the student from books and periodicals
3 Solutions to set exercises or problems

4 Test papers

5 Oral presentation of individual work

6 Individual work of a practical nature, e.g. making a film loop or
   model (Personal communication 1969)

It will be noticed that these proposals are very wide in scope, and in
particular that (3) and (4) would already be present in any conventional course. (This was a compromise position, in contrast to A.T.M.
schools project where the tendency had been to want to move almost
exclusively to activity represented by the alternative (1) above.) This
implied that a course could be devised in which conventional lecture
courses, assessed by the usual end of course test, presenting material
judged essential to a student's progress, were balanced by opportunities for individual investigational work, the importance of which
could be underlined by the attitudes and enthusiasm of tutors for this
approach.

The outcome was extremely interesting; a great deal depended on
the care with which the students were introduced to the underlying
ideas, and on the degree of commitment felt by tutors to the whole
enterprise. Where a weak student started off on the wrong foot with
an unsuitable topic, he was in danger of forming a view of the nature
of mathematics, and of the purpose of the studies, which one felt was
unfortunate – as when tedious enumerations of data were assembled
without recognition of existing patterns, or conjectures were stated,
on dubious evidence, with little realisation of the significance of
proof. Yet many students produced work which could only be
described as truly mathematical in character, in that the processes
of mathematics (classifying, recognising analogies, formulating conjectures, devising notations or symbolism, constructing counter-
examples, refutations and proofs) were all explicitly demonstrated.
(See also Branfield (1969) for an example of these processes in action.) Some examples of investigation topics are given on p. 118.

One particularly fruitful exercise attempted by some students was
that of 're-visiting', as third year students, an investigation which
had been attempted in the first year. On the second reading the purpose was not so much to extend the investigation of the original

problem as to analyse the mathematical processes which one had used in attacking it. ('Here I was exploring . . . now making a conjecture . . . tested it . . . then I decided on a notation . . . tried to prove my statement . . . modified the conjecture.')

Perhaps the most difficult feature of this approach was maintaining the appropriate balance of intervention and control over students' 'independent' work. Too much tutor intervention nullifies the whole exercise, yet it seemed that without fairly firm guidance (which might in part come from collective class discussions of problem-solving, especially in the early stages), students could readily stray into unprofitable approaches. There was also a tension, perhaps inevitable, between the demands of course activities as a learning experience – in which tutor help might be entirely appropriate – and the requirements of the assessment procedures in which it seemed necessary to be clear on the student's own contribution to the work. To some extent, the tutors' personal involvement in the situation, and the possibility of an 'oral' with the external examiner, overcame this difficulty, yet the fact remains that the only thing a moderator can assess is the work that is there at the end of the course. It is sometimes difficult, too, to distinguish 'from outside' how profitable has been a student's independent reading of some well-established piece of mathematics (pedestrian reproduction of standard texts was occasionally the response of weaker students, if unchecked), though again there is evident value in training students to extract for themselves the content of books and periodicals. The problems of assessment and of syllabus coverage may have been part of the reason it was felt necessary to revert to more traditional means of assessment for the fourth year of the B.Ed. course.

Conclusions

On balance this approach seems to have much to offer.

. . . at its best, the method has many advantages, in particular those of reducing the pressure of assessment procedures on the form and content of the course and allowing students to work independently in a wider range of mathematical activities, rather than receiving mathematics for subsequent regurgitation. (Examiner's Report, 1973)

A further discussion of some of the difficulties, of teaching and of assessing, is given in Eagle and Watson 1972, from which the following is an extract:

The difficulties of testing problem-solving ability are considerable but not insuperable. It seems worth devoting effort to this aspect of assessment so that in our teaching we can maintain a balance between the 'content' and process aspects of mathematics.

The absorption of the majority of colleges into the structure of the polytechnics, with a consequential reorganisation of mathematics courses, is bound to change existing patterns, though whether the investigational approach will flourish or wither in its new surroundings it is too early to say. It is certainly the case that many of the teachers who experienced such a course have found it natural to develop a similar teaching approach with their own pupils, and some others who have encountered the ideas through in-service courses or at A.T.M. conferences have embraced 'investigations' with enthusiasm.

Moreover, the portended changes in the pattern of sixth form teaching – with the advent of C.E.E., the discussions on N and F, and of a common examination at sixteen plus, may give opportunities for the encouragement of this approach in schools. It is noteworthy that among the seven Schools Council exploratory studies is one 'emphasising the role of individual investigation' to be conducted by the A.T.M. and based at the Shell Centre for Mathematical Education, Nottingham University (Quadling 1975).

Appendix

A few examples of investigations used with college students
(B. J. Lang, personal communication, 1975)

1 On a square lattice of dots, polygons are drawn whose vertices are lattice points. Investigate the relationship between the areas of the polygons and the number of lattice points lying (a) inside the polygon, and (b) on the boundary of the polygon

2 Investigate recurring decimals

3 How many shortest paths are there between two points on a rectangular grid?

4 Investigate Pythagorean triples, i.e. sets of three integers x, y, z, such that $x^2 + y^2 = z^2$

5 'In any triangle the sum of the three . . . is greater than the semi-perimeter.'

Investigate the truth of the above statement when the words (a) altitudes, (b) medians, (c) angle-bisectors, are inserted

PART III

14 Where next? Some of the problems

Why mathematics?

Mathematics had its origins in practical matters (surveying, navigation, commerce) and acquired its place in the curriculum of schools in Britain as a response to changes in the requirements of society. These changes continue, and unless we acknowledge them we may find ourselves teaching for a bygone age (see 'The Saber-tooth Curriculum' by H. Benjamin in Hooper (1971) pp. 7–15). Though we no longer require armies of clerical workers with copperplate and impeccable arithmetic (*vide* typewriters and computers), society does require of everyman a basic competence in mathematics, misleadingly called 'numeracy' (and usually left undefined in discussion, with confusing results!). Handling money, interpreting graphs and charts, thinking logically are useful skills for anyone. Society has also certain requirements of skilled manpower — research mathematicians who will advance the subject; teachers who will pass on the lore of mathematics; scientists and engineers, managers and businessmen who will *use* mathematics at varying levels of sophistication; craftsmen, technicians, and others whose requirements will be related to their work.

Controversy has developed recently in connection with the last-named group. It has been suggested that the requirements of industry are not being met, children being deficient in elementary arithmetic or lacking facility in simple algebra. The difficulties are blamed on 'modern mathematics', and though they are *not* new, they cannot be ignored. There is a clear gap between schools and industry, neither side knowing very much about what happens in the other half. Precisely what the requirements of employers *are* is not always clear; thus a complaint 'trainees cannot multiply $1,240 \times 680$ without logarithms' prompts a query about calculators and how such a computation is done on the shopfloor, though fears are expressed that the use of calculators in schools may make matters worse. (An endearing story is that of the interviewee who, asked how many sixty-fourths were contained in one inch, answered with engaging truthfulness that he didn't know but he thought 'there must be an awful lot of them'.) Schools have often emphasised method and problem-solving skills rather than the importance of computational accuracy; neglect of basic practice has resulted, and pupils sometimes show a cavalier disregard for the appropriateness of 'real-life' answers. The consequences of ordering 730 tons of sheet steel instead of the correct 73 tons are plain to see, whereas a calculation which 'merely' has the decimal point in the wrong place may seem 'near enough' correct. The re-written S.M.P. main course included additional revision materials. Manipulative skills have since received attention (School Mathematics Project 1974) and several conferences of the Institute of Mathematics and its Applications on the school/employment interface have been held (see, for example, Tammadge 1975).

Although the requirements of society and of industry are ample justification for the place of mathematics in the curriculum, the goals they imply are by no means the only ones for mathematics teaching. No mention has been made yet of the requirements of *pupils*. These other goals have their own influence on the way mathematics is taught. Since the whole world about us, natural and man-made, has mathematics deeply embedded in it, we owe it to our pupils to help them to use mathematics to comprehend the world, to see how men in general and they in particular can use intellect to understand and shape environment. 'Mathematics develops patterns of thinking which are fundamental patterns of all thinking' (Mathematical Association 1959, p. 2). As an example of deductive thought it makes a distinctive contribution to personal development and many of its concepts (impossibility proofs, optimisation, isomorphism, conjecture, generalisation) have a wide relevance in human life.

Mathematics is traditionally regarded as unpopular, so that 'enjoyment' cannot be a compelling reason for teaching it. Yet where it is approached sensitively, with credit given for effort, involvement, reasonable conjecture as well as *correctness* (for we are dealing with immature 'mathematicians' who will easily be discouraged by lack of success), pupils *can* find pleasure and excitement in their work. Children usually enjoy a puzzle, but no-one likes being publicly humiliated, so that many withdraw themselves from mathematics as it is sometimes taught.

Reynolds (1974) writes 'a society that has assumed English and mathematics to have a basic and essential position may be fading away' and goes on to enquire whether we are witnessing the dawn of a 'mathematical dark-age' in our schools. 'The traditional mathematics syllabus (should) be regarded very critically and the purpose for which any of it is included should be reviewed' (Mathematical Association 1959, p. 3). Perhaps we should look equally critically at our reasons for teaching mathematics and how these are reflected in our methods and approaches.

What is mathematics?

'Arithmetic — money, weight, measurement, decimals, income tax, . . .'
'the language of science'
'anything under 510 in the Dewey classification'
'logical inference; proof'
'science of space and number' (*The Little Oxford English Dictionary*, 1941)
'the study of all possible patterns' (Sawyer 1955)
'the study of abstract structure'
'queen and servant of science'

All these quotations, real or imagined, describe some aspect of mathematics. What we think it *is* determines how we teach. The product/process dichotomy has already been discussed in chapter 13 where the view of mathematics as solely a body of knowledge was questioned. Although the 'process' approach to mathematics currently receives much less attention than it merits, it can, if carried to extremes, be harmful to pupils, whose limited grasp of existing mathematical knowledge hampers the very problem-solving work on which they have concentrated.

'Mathematics is a language', though true of its use of symbolism, is dangerously misleading if the problem of understanding it is seen only as one of *translation*, for *concepts* are involved also. (A book on chess, written in French, might be unintelligible because one knew nothing of chess, or nothing of French, or both!). Hence the failure, all too often, of 'understand mathematics in twelve lessons'; there is no 'royal road'.

Skemp (1971) describes mathematics as a complex structure of ideas, organised in a hierarchy, a useful way of viewing the way the more sophisticated ideas are based upon and related to simpler ones in lower 'layers', which may be pre-requisites to understanding. Skemp is concerned here with psychological structure; some pure mathematicians see mathematics as a study of structure in a different sense – that exemplified by the ideas of group, field, vector, space, . . .

The pupil must be taught to be aware of the absolute necessity of an axiomatic approach to mathematics. At the earliest possible moment, the pupil must get used to constant dealings with abstract concepts, one of the most difficult of which is the concept of a *mapping* . . . since in teaching these two points, we are dealing with the very *cornerstones* of the structure of modern mathematics, everything else taught during these early years ought to be made to *take second place*. (Dieudonné 1969, p. 13)

The B.B.C. television series *Maths Today*, intended for the first two years of secondary school, adopted an approach based upon abstract structure – so that, for example, fractions (rationals) were considered as equivalence classes of ordered pairs of integers – although plenty of concrete embodiments and realisations of these were provided. It is doubtful whether this approach is suitable for most children in schools and it has not generally commended itself in this country.

Whether mathematics should be studied as 'pure' or 'applied' (queen or servant of science) is a debate of long-standing. We have already considered (chapter 12) the latest attempts to reconcile the two approaches and have not space to discuss the matter further (see Bickley 1958; Allanson 1968; Hammersley 1968). It is evident that the two strands must somehow be brought together in teaching, and that this has implications for examinations, too.

The activity of 'mathematising' a given situation (an ugly word!), or imposing a pattern or structure of our *own* choosing is an alternative to a study of the 'important structures' of mathematics; this approach, already mentioned in chapter 13, is intuitive and centred upon the pupil rather than the mathematics. Rather than attempting to transmit known mathematics we are concerned with the creative side of the child's learning, trying to help him to build mathematics for himself. Several interesting examples are given in Banwell and others (1972) and in numerous issues of *Mathematics Teaching*.

Links with other subjects

The changes in mathematics and science curriculum in the early sixties tended to occur independently of each other; each group of reformers was generally pre-occupied with the internal restructuring of the subject. But scientists soon began to express their disquiet, for example at the use of the slide rule instead of logarithms (which were initially excluded from the S.M.P. O level syllabus), and because pupils were unable to 'change the subject' of a formula such as $l_t = l_0 (1 + \alpha t)$. Admittedly less emphasis was placed on manipulative ability in the new approach (and that this had been carried too far was acknowledged later–School Mathematics Project 1974). Yet there have always been pupils whose mathematical skills were inadequate for their work in science. Now 'modern mathematics' provided a convenient scapegoat! The logarithm/slide rule controversy is fast becoming irrelevant with the advent of the electronic calculator. It is possible that teachers were comparing their pupils' abilities with their own (probably above average) proficiency at a similar age; the important issue is to discover what are the *essential* mathematical skills required at any stage in science. (The picture is complicated further by the recent spread of mathematics into a range of other subject areas, such as geography (Cole and Beynon 1969), economics, and so on. It is difficult to meet the sometimes conflicting requirements of the 'users'.) In a brilliant satire containing some uncomfortable home truths, Allanson (1968) states: 'For a scientist, whether natural or social, a capacity for using mathematical models and arguments is a necessity. In brief, mathematics has now become so important that we can no longer leave it to the mathematicians.' Recriminations and the scoring of debating points do not help the pupil, however, and after the initial controversies groups of scientists and mathematicians sat down to the more exacting task of seeking solutions (Royal Society 1973, 1974; British Committee on Chemical Education 1974; see also Lewis 1973 for the social sciences).

Although the identification of appropriate syllabus content is one aspect of the problem, a more important condition for progress seems to be dialogue between teachers within the school. Even mutual examination of textbooks, work sheets, and test papers can do much to overcome difficulties of nomenclature, syllabus sequence, and amount of drill and practice; the attitude of the teachers may be a much more significant factor than any structural change (Bacon and Webb 1975).

The Continuing Mathematics Project at the University of Sussex has prepared 'self-learning materials in mathematics for post O level students of biology, geography, and economics'. These include packages combining written and audio-visual materials on introductory calculus, statistics and probability, critical path analysis, and a variety of basic 'manipulative' topics. Their design does not assume the availability of a specialist mathematics teacher, so that they may be particularly useful in relieving the present shortage of specialists.

Computers and calculators

Although generally associated with mathematics, computer studies have a border line position. In Scotland where systematic support has been given from central resources, the work is generally integrated into mathematics and business studies. In England and Wales a more haphazard picture emerges. A few schools have a separate department of computer studies and there are A level and C.S.E. examinations available, with some O levels recently announced. Efforts are being made to widen out by involving teachers of other subjects (geography, commerce, science . . .), but to date most teaching has been done by interested mathematicians.

Ideas from digital computing have influenced the 'modern' syllabuses, leading to the introduction of work on flow-charting or programming and at A level some topics of numerical analysis (M.E.I., S.M.P., Scottish Certificate of Sixth Year Studies), and, for younger pupils, on errors, Boolean algebra, and binary arithmetic. Work on analogue computers has been less widespread, though their use in the study of differential equations of first and second order can provide an interesting link with physics.

Access to computers, though becoming increasingly easier, is still something of a problem, so that even teachers who wish to make more use of the computer are not able to do so. The situation with regard to electronic pocket calculators, however, is very different. These have spread so rapidly since their introduction in 1971 that mathematics teachers are having to give thought to their use in the classroom. Apart from their value for such topics as statistics, iterative processes, and error-estimation, they allow the use of realistic (or actual) data in numerical work. No longer need triangles have sides 3, 4, and 5 metres, and the peculiar preference given to circles of radii 7, 14, 21, etc. is at an end! This development has led to some disquiet – will children lose whatever computational facility they now possess (compare the predictions of a race of legless motorists), and can calculators be permitted in examinations? (They are generally not permitted in C.S.E. and O level mathematics examinations, partly because of possible 'unfairness', partly because in any case it is desired to test simple computational ability. In fact a calculator would be of little help in most current examination questions. These rulings are largely the decisions of teacher panels, and the variability in regulations reflects the lack of unanimity among teachers on this matter.)

Another important issue is how the teaching of mathematics might be modified to take advantage of the availability of calculators (see Ellis and others 1971; Edsall 1976). As an example a solution to $x^3 = 10x+1$ can be found by trial (or, for older pupils, by an iteration) as quickly as $x^2 = 10x+1$ could be solved by formula and tables. The calculator may also be used in investigating patterns, such as the following:

'$(n^3 - n)$ is always divisible by 3 for any positive integer n.'
Is this true? If so, try to prove it.
Now investigate $(n^5 - n)$ and $(n^7 - n)$.

Teaching methods

The emphasis on understanding, discovery, pattern, structure, which characterised the 'modern' developments led, as we have seen, to some neglect of manipulative and computational facility. The role of techniques and practice was perhaps underestimated. The reaction against 'mindless repetition of half-understood processes' was understandable. But although music is a trumpet solo or a Beethoven symphony rather than scales and five-finger exercises, the latter represent the practice which is essential if any reasonable standard of performance, and the performer's enjoyment in it, is to be reached. Though written at the outset of the current wave of reform, the words which follow have not always been heeded. 'Practice without the power of mathematical thinking leads nowhere; the power of mathematical thinking without practice is like knowing what to do but not having the skill or tools to do it' (Mathematical Association 1955, p. 4).

Mathematics teaching now involves much more practical activity than formerly, making desirable the provision of specialist accommodation and storage facilities for the array of geoboards, centimetre cubes, graph- and grid-sheets, surveying equipment, mosaic tiles, dice, Dienes blocks, work cards, students' project folders, and so on, which may nowadays be in use. Desks, textbooks, blackboard, chalk, and an answer book are no longer all that need be provided! Though many schools have inadequate facilities these considerations are increasingly borne in mind when buildings are planned or modified.

A wide variety of aids is also in use. The overhead projector has great potential (see Backhouse and others 1974) and television programmes are widely used, particularly in primary schools: a survey found that 50 per cent of primary schools used the B.B.C. programme *Maths Workshop* (Schools Council 1974). Computer-assisted instruction is still largely at the stage of research, although the Hertfordshire Computer Managed Mathematics Project uses the computer to mark tests, and monitor childrens' progress through a network of work sheets, and teaching systems using the computer are being developed in higher education (Hooper 1975; see also Holmes 1975).

New organisational patterns of teaching emanating for the most part from other curriculum areas have had their impact on mathematics teaching. Team teaching within the mathematics department has many potential advantages; sharing the burden of preparation of materials, effectively increasing the teacher/pupil ratio and the availability of particular expertise, and providing a mechanism for sharing experience and 'inducting' inexperienced teachers. It does, however, make heavy demands, on time for consultation and planning, and on tact and sensitivity!

'Integration' (in the curriculum sense) is much in the air at present, and the place of mathematics as a distinct curriculum subject was an issue raised by the 'continuation' project (chapter 11). It is true that the links between mathematics and other subjects have not been explored very energetically in the past, and that much more could be done. But this is not to say that mathematics should be completely submerged, entirely losing its separate identity, for there is the danger that then the links *within* mathematics would be neglected. The explanatory power of mathematics resides largely in the way in which many different situations are recognised to be similar *when viewed mathematically*. It seems probable that the coherence and cross-linking which are essential to mathematical understanding necessitate *some* separate consideration of mathematics as such. Nevertheless, despite the additional demands this must make, it seems desirable that mathematics teachers should explore the opportunities for collaboration which do arise.

Team-teaching is unlikely to be imposed upon mathematics teachers. But this is not so of mixed ability groupings. Although it could not be claimed that the sample was representative, a recent survey of non-selective schools indicated that mixed ability teaching had been introduced 'by decree of the head' in twenty-four cases out of sixty-two, often without any prior consultation with staff (Assistant Masters Association 1973, pp. 8, 9), and noted that 'mathematics teachers in secondary schools were more reluctant than other specialists to adopt their approach and methodology to the needs of mixed ability groups' (ibid., p. 24). Some teachers have adopted this method and found it very successful, others remain antagonistic or just cautious (Kennard and others 1974; Brissenden 1973). There may be special reasons, unrelated to a preference for the *status quo*, in the case of mathematics. Admittedly, its 'social' aspect (discussion, collaboration in working, argument, agreement on definitions) has been neglected in the past. But mathematics has a personal aspect, too. 'No one else can *understand* it for you.' The group or individual working which mixed ability teaching seems to demand may result in some children being 'passengers'. (One may watch and enjoy a play in an unknown language, but in mathematics the 'words of the play' are important. Even if poetry, music, art, history may be taught successfully in mixed ability groups it does not follow that mathematics can; nor that it cannot). If the teacher can successfully involve all the children, well and good but this may prove beyond his powers, given the limitations under which he must work. If teachers are placed in a situation to which they cannot adapt, the social benefits of mixed ability grouping may be outweighed by the losses in other directions, and the last state be worse than the first. This is not to imply that mixed ability teaching should be resisted at all costs, but one hopes that teachers will be willing to experiment with it, and that heads will avoid introducing it by decree!

Teacher education

Despite all publicity and the efforts of the past fifteen years mathematics teachers remain in short supply. The outlook is gloomy. Colleges of education, after a period of expansion in which much thought was given to mathematics courses, have been thrown into the organisational melting pot. Student numbers in mathematics in some colleges have been very low; a cutback in total numbers may hit mathematics harder than many subjects. Although university output of mathematicians has increased considerably many graduates have gone into industry or higher education. On the credit side, the two Shell Mathematics Centres have opened, the number of mathematics tutors in university departments of education has doubled, and local education authorities have appointed mathematics advisers and created teachers' centres. The Mathematics Teacher Education Project should also be mentioned. This was set up to develop a bank of resource materials for the training of graduate mathematics teachers – see Wain and Woodrow (1973), and cf. the parallel science project, Sutton and Haysom (1974).

There have been encouraging developments in in-service training, too, although the 'glorious vision' of the James Report and the White Paper which followed it await a more favourable financial situation.

The Department of Education and Science, universities, local education authorities, and many other agencies (e.g. S.M.P. itself; Thwaites 1972, p. 245) have provided courses and conferences, and a network of activities, in teachers' centres and elsewhere, has sprung up. How this can be sustained and encouraged is one of the major problems; in my view progress will be very slow until a situation is reached when in-service updating is freely accepted by teachers as part of their conditions of service. Apart from 'official' provision there are plenty of opportunities for teachers who wish to keep up to date; the Mathematical Association and the Association of Teachers of Mathematics, their conferences, reports and journals. (The latter are usually available in teachers' centre and other libraries, as well as through school or personal subscription.) There are also numerous diploma and degree courses (and even schools television broadcasts during the holidays!).

Examinations

Some recent developments in techniques of assessment have been referred to above; the articles by Wood and by Bishop in Association of Teachers of Mathematics 1968, and by Gray in Sutton and Haysom 1974, are also worth reading. The pressures exerted by examinations on the curriculum (the so-called backwash effect) sometimes result in unwilling pupils being forced to work through a constricting external syllabus because of parental, employer, or societal pressures. I believe it is unrealistic to seek the abolition of examinations (Griffiths and Howson (1974) describe how they were instituted to replace the system of patronage), and that the solution lies in the professional independence of teachers constantly alert to act in the best interests of their pupils, by using their control over the school curriculum, and their involvement in the assessment process.

Major changes in the structure of examinations at sixteen plus and eighteen plus are impending. The form to be taken by N and F levels, and the *feasibility* studies of a common examination at sixteen plus (rather pre-empted by the Schools Council's approach of 'decide first, examine the likely consequences afterwards'?) are currently the subject of considerable debate. The articles by Hersee (1974) and Brown (1975) give useful information.

Amalgamation of C.S.E. and G.C.E. would have advantages in reducing the load on some pupils and the length of the examination 'season' and might improve the image of C.S.E. But it would create many other problems, too, and it is not yet clear where the balance of advantages lies. At N/F level it is encouraging that the mathematics steering group of Schools Council is examining no fewer than seven possibilities for curriculum in mathematics and statistics (Quadling 1975). One question which must be asked of any future arrangements is whether they will assist or inhibit the sort of initiatives in curriculum reform which S.M.P. and the A.T.M. project were able to make with the cooperation of the examining boards.

Future developments

It would be pleasant to believe that research in mathematical education is on the point of providing guidelines for action in curriculum change. Though I do not subscribe to the view that research is inherently incapable of supplying the answers we need (or, at least, some of them), the situation is not that simple. Even where research findings do have clear classroom implications, there remains the problem of communicating them to teachers, which some have suggested is best solved by involving teachers themselves in the research. In terms of getting things done there is no clear line between 'research' and 'development'; teachers who are 'involved' in an investigation may benefit as much from the by-products (greater awareness, seeing an alternative point of view, self examination . . .) as from the enquiry itself. A warning is necessary, however; although teaching is undoubtedly an art, we must also work to make it a *science*. Basic research (for example, that on learning) can tell us something of value. Without it we may find ourselves at the mercy of received opinion or current fashion. Attempts to evaluate curricula are only just beginning. This is an area where we have a lot to learn.

No one can answer the question, 'Is the S.M.P. course better than those it replaced' (if, indeed, the question has any meaning!) and evaluation has generally been 'formative', attempting to improve the effectiveness of course materials for the job they were designed to do (see Schools Council 1973 for several case studies, including two on mathematics).

Conclusion

To anyone who expected to find here a blueprint for teaching mathematics, I apologise! No universal formula will guarantee success; pupils, teacher, and circumstances all affect the outcome. This view has major implications for the control and planning of education — for no amount of high level 'thinking out the right answer' will get us where we want to be. This is not to argue for anarchy; there have to be some limitations on individual freedom of action, and common policies and collaborative action have their advantages. But the local autonomy which teachers presently enjoy *allows the possibility of evolutionary change* and should not be lightly discarded.

There are no easy solutions to the problems raised in this book; otherwise they would have been found by now. Impossible dilemma? Creative tension? — Whatever we call it, we have to try to keep a balance between the forces pulling in different directions, constantly re-assessing our aims and methods, and responsive to changing situations. The teaching of mathematics is not a static craft which once learned becomes routine; indeed a feeling of settled security, with all the teacher's problems 'taped' may well be a danger signal!

I am convinced that material covered is much less important than the spirit in which it is approached. When talking recently with not unintelligent children I have been saddened, not by the fact that they didn't understand mathematics, but that they had *stopped expecting to understand it*. French irregular verbs, whatever their logic, have to be learned because that's the way French people speak. In mathematics the pupil should be able to decide for himself, quite independently of the teacher, the correctness and value of his work, for the basis of mathematics is rationality and conviction by *proof*, not by appeal to authority.

What it comes down to, after all, is this. 'Learning by discovery', 'algebraic structure', 'computational facility', 'investigating a situation' — do they help us to answer the question, 'what is mathematics in the curriculum *for*'?